Human Factors in Intelligent Vehicles

RIVER PUBLISHERS SERIES IN TRANSPORT TECHNOLOGY

Indexing: all books published in this series are submitted to the Web of Science Book Citation Index (BkCI), to SCOPUS, to CrossRef and to Google Scholar for evaluation and indexing

The "River Publishers Series in Transport Technology" is a series of comprehensive academic and professional books which focus on theory and applications in the various disciplines within Transport Technology, namely Automotive and Aerospace. The series will serve as a multi-disciplinary resource linking Transport Technology with society. The book series fulfils the rapidly growing worldwide interest in these areas.

Books published in the series include research monographs, edited volumes, handbooks and textbooks. The books provide professionals, researchers, educators, and advanced students in the field with an invaluable insight into the latest research and developments.

Topics covered in the series include, but are by no means restricted to the following:

- Automotive
- Aerodynamics
- Aerospace Engineering
- Aeronautics
- Multifunctional Materials
- Structural Mechanics

For a list of other books in this series, visit www.riverpublishers.com

Human Factors in Intelligent Vehicles

Editors

Cristina Olaverri-Monreal

Chair ITS-Sustainable Transport Logistics 4.0
Johannes Kepler University Linz
Austria

Fernando García-Fernández

University Carlos III de Madrid
Spain

Rosaldo J. F. Rossetti

Faculdade de Engenharia da Universidade do Porto
Portugal

River Publishers

Published, sold and distributed by:
River Publishers
Alsbjergvej 10
9260 Gistrup
Denmark

www.riverpublishers.com

ISBN: 978-87-7022-204-4 (Hardback)
978-87-7022-203-7 (Ebook)

Contents

Preface

Intelligent Vehicle technologies have experienced a great evolution/advancement in the last couple of decades, turning vehicles themselves into progressively more interactive elements in transportation and mobility systems.

In addition, during the last few years, significant attention has been paid to developing and implementing key technologies of future Intelligent Transportation Systems (ITS) by integrating vehicular electronics that enhance road safety and prevent traffic accidents. The fields of sensing, communication and control technologies play an increasingly crucial role for vehicle safety and security, while research in transport continues among industry engineers, practitioners, students and government agencies. However, analyzing the impact of such technologies on driver traffic awareness and behavior, specifically towards improving driving performance and reducing road accidents, still demands improved tools and approaches.

While the feasibility of incorporating new technology-driven functionality to vehicles has played a central role in automotive design, safety issues related to interaction with the new in-vehicle systems have not always been taken into consideration. A system that guarantees efficiency of use, comfort and user satisfaction can contribute to a more conscious driving behavior that would directly result from the adoption of intelligent vehicle technologies.

The collection of work presented in "Human Factors in Intelligent Vehicles" (HFIV) aims to address issues related to the analysis of human factors in the design and evaluation of intelligent vehicles for a wide spectrum of applications and over different dimensions. To commemorate the 9th anniversary of the IEEE ITS Workshop on Human Factors (http://hfiv.net), that is promoted by the IEEE ITS Society's Human Factors in Intelligent Transportation Systems (HFITS) Technical Activities Sub-Committee, some recent works of authors active in the automotive human factors community have been gathered in this River Publishers book with the collaboration of the Artificial Transportation Systems and Simulation (ATSS) Technical Activities Sub-Committee.

The work presented in this collection is expected to disseminate knowledge among the technical and scientific communities, practitioners and students alike, and contribute to the development and enhancement of state-of-the-art approaches.

This book serves as a platform for scientific knowledge exchange and experience-sharing. Enclosed here are extended versions of manuscripts that were presented at the IEEE ITSS Workshop on "Human Factors in Intelligent Vehicles" and that contain additional developments in terms of deeper analysis and detailed experimental/simulation results.

The various authors examine autonomous vehicles as well as the frameworks for analyzing automation, modelling and methods for road users' interaction such as intelligent user interfaces, including brain-computer interfaces and simulation and analysis tools related to human factors.

The book is divided in the following seven chapters:

Chapter 1 "Continuous Game Theory Pedestrian Modelling Method for Autonomous Vehicles", co-authored by several researchers from Ibex Automation Ltd and the universities of Leeds, Lincoln, and De Montfort in the United Kingdom elaborates on game-theoretic predictive parameters. The work shows how these parameters can be applied to pedestrians' natural and continuous motion during road-crossings to make predictions about their interactions with the controllers of autonomous vehicles (AV) in a variety of real-world settings.

Trust and perceived workload with regard to partially automated vehicles and its correlation with information presentation in the vehicle is the topic of Chapter 2, "The interface challenge for partially automated vehicles: how driver characteristics affect information usage over time". The authors from the Universities of Warwick and Coventry in cooperation with Jaguar Land Rover in the United Kingdom contribute to future human-machine-interaction (HMI) design through a better understanding of how driver characteristics can affect information use inside partially automated vehicles.

In Chapter 3 "A CNN Approach for Bi-Directional Brainwave Controller for Intelligent Vehicles" brain-computer interfaces (BCI) are investigated as frameworks to enable severely disabled people to drive vehicles. The authors from the University Carlos III in Madrid, Spain review some of the most sophisticated and relevant techniques in the classification of brain patterns and propose a new method by using a low-cost helmet and convolutional neural networks (CNN). They showed the feasibility of their approach by improving prediction results.

Chapter 4 titled "A-RCRAFT Framework for Analysing Automation: Application to SAE J3016 Levels of Driving Automation" by authors from ICS-IRIT at Toulouse University (France) analyses automation by delineating some key concepts that are emphasized separately in the literature of automation design. This A-RCRAFT framework provides an analytical support structure with the concepts of Allocation of Resources, Control Transitions, Responsibility, Authority, and Functions and Tasks (A-RCRAFT). Examples of how the framework can be used to analyse different options of driving automation design according to the SAE J3016 levels of driving automation are shown.

Autonomous vehicles (AV) and their impact on trust and pedestrian behavior as well as the necessity of communication protocols are the topics of Chapter 5 "Autonomous Vehicles: Vulnerable Road User Response to Visual Information Using an Analysis Framework for Shared Spaces". The authors from the Johannes Kepler University at Linz in Austria studied the effects of different communication paradigms in shared spaces and developed an algorithm for the analysis of pedestrian behavioral patterns such as the pose and distance of the pedestrian.

"Intelligent Vehicles and Older Drivers" is the topic of Chapter 6. The authors of the chapter, from the HumanLAB, DGIST (Daegu Gyeongbuk Institute of Science and Technology) and Technical Center, Sonnet.AI in South Korea elucidate the different human physiological resources involved in driving tasks such as vision, hearing, cognition, and physical function and propose intelligent-vehicle-based solutions that might compensate for the reduction of these physical capabilities due to advancing age.

Finally, Chapter 7 "Integration model of multi-agent architectures for data fusion-based active driving systems" presents an architecture model to integrate new active safety systems into Human-In-the-Loop (HITL) driving simulators. The authors from the Computer Science Department at the Universidad Carlos III in Madrid, Spain gathered data from several information providers. A support system merged a variety of driver parameters with the corresponding driving scene through a data fusion process, allowing this procedure for a more realistic system performance for measuring human factors elements in a driving context.

Much appreciation is owed to all of the authors that have submitted their contributions to this book and for sharing their novel visions, outstanding research and significant results. The chapters included in this book have benefited greatly from the laborious and time-consuming work of many anonymous reviewers that have offered their expertise, suggestions and recommendations, and we therefore wish to thank them as well. We hope that the readers will enjoy the content of this book as much as the editors did, and that the scientific community and practitioners alike will find this work stimulating and useful in promoting and developing the field of Human Factors in Intelligent Vehicles.

Cristina Olaverri-Monreal
Fernando García-Fernández
Rosaldo Rossetti

List of Contributors

Agapito Ledezma, *Departamento de Informática, Universidad Carlos III de Madrid; E-mail: ledezma@inf.uc3m.es*

Araceli Sanchis, *Departamento de Informática, Universidad Carlos III de Madrid; E-mail: masm@inf.uc3m.es*

Armando Astudillo Olalla, *Intelligent Systems Lab – University Carlos III Madrid, Spain; E-mail: aastudil@ing.uc3m.es*

Arun Ulahannan, *Coventry University, United Kingdom; E-mail: Arun.Ulahannan@coventry.ac.uk*

Charles W. Fox, *Institute for Transport Studies, University of Leeds, UK; School of Computer Science, University of Lincoln, UK; Ibex Automation Ltd., UK; E-mail: chfox@lincoln.ac.uk*

Cristina Olaverri-Monreal, *Chair ITS-Sustainable Transport Logistics 4.0; Johannes Kepler University Linz, Austria; E-mail: cristina.olaverri-monreal@jku.at*

Fanta Camara, *Institute for Transport Studies, University of Leeds, UK; School of Computer Science, University of Lincoln, UK; E-mail: tsfc@leeds.ac.uk*

Fernando Garcıa Fernandez, *Intelligent Systems Lab – University Carlos III Madrid, Spain; E-mail: fegarcia@ing.uc3m.es*

Joonwoo Son, *HumanLAB, DGIST (Daegu Gyeongbuk Institute of Science and Technology), 42988, Daegu, Technojungang-daero, South Korea; Techical Center, Sonnet.AI, 06764, Seoul, Taebo-ro, South Korea; E-mail: joonwooson@gmail.com*

Myoungouk Park, *HumanLAB, DGIST(Daegu Gyeongbuk Institute of Science and Technology), 42988, Daegu, Technojungang-daero, South Korea*

Natasha Merat, *Institute for Transport Studies, University of Leeds, UK; E-mail: n.merat@its.leeds.ac.uk*

Nicola Bellotto, *School of Computer Science, University of Lincoln, UK; E-mail: nbellotto@lincoln.ac.uk*

Oscar Sipele, *Departamento de Informática, Universidad Carlos III de Madrid; E-mail: bsipele@inf.uc3m.es*

Paul Jennings, *WMG, University of Warwick, United Kingdom; E-mail: Paul.Jennings@warwick.ac.uk*

Serhan Cosar, *Institute of Engineering Sciences, De Montfort University, UK; E-mail: serhan.cosar@dmu.ac.uk*

Simon Thompson, *Jaguar Land Rover; E-mail: sthom261@jaguarlandrover.com*

Stewart Birrell, *Coventry University, E-mail: S.Birrell@coventry.ac.uk*

Walter Morales Alvarez, *Chair ITS-Sustainable Transport Logistics 4.0; Johannes Kepler University Linz, Austria; E-mail: walter.morales alvarez@jku.at*

List of Figures

List of Tables

List of Abbreviations

1D	one-dimensional
2D	two-dimensional
3D	three-dimensional
AAR	Adaptive Auto-Regressive
ACC	Adaptive Cruise Control
ADAS	Advanced Driving Assistance System
ADS	Automated Driving Systems
A-RCRAFT	Allocation of Resources, Control transitions, Responsibility, Authority, Functions and Tasks
AFS	Adaptive Front-lighting System
AV(s)	Autonomous Vehicle(s)
BCI	Brain–Computer Interfaces
BMK	Austrian Ministry for Climate Action, Environment, Energy, Mobility, Innovation and Technology
BSD	Blind Spot Detection
DALI	Driver Activity Load Index
CC	Creative Commons
CNN	Convolutional Neuronal Network
CNS	Car Navigation System
DARPA	Department Advanced Research Projects Agency
DDT	Dynamic Driving Task
DTW	Dynamic Time Warping
DWT	Discrete Wavelet Transformation
ECG	Electrocardiogram
EEG	Electroencephalogram
eHMI	external Human–Machine Interface
ETL	Extract-TransformLoad
FCW	Forward Collision Warning
DBQ	Driver Behavior Questionnaire
FCWC	Forward Collision Warning Counts
FFT	Fast Fourier Transform

FWS	Flight Warning System
GMOC	Goal, Model, Observability, and Controllability
HCI	Human–Computer Interaction
HMI	Human–Machine Interface
HTIL	Human-In-the-Loop
HTTP	Hypertext Transfer Protocol
iCab	Intelligent Campus Automobile
ID	Identifier
ITS	Intelligent Transport Systems
IV	Intelligent Vehicles
JAAD	Joint Attention for Autonomous Driving
JADE	Java Agent Development Environment
JPDAF	Joint Probabilistic Data Association Filter
LDA	Linear Discriminant Analysis
LDA	Levels of Driving Automation
LDW	Lane Departure Warning
LDWC	Lane Departure Warning Counts
LEDs	Light-EmittingDiodes
LiDAR	Light Imaging Detection and Ranging
LoA	Levels of Automation
ISI	Intelligent Systems Laboratory
MHT	Multiple Hypothesis Tracking
NHTSA	National Highway Traffic Safety Administration
NLLLoss	Negative Log Likelihood Loss
NN	Neural Network
NN	Nearest Neighbor
NV	Night Vision
OEDR	Object and Event Detection and Response
ODD	Operational Design Domain
PAFs	Part Affinity Fields
PDCA	Plan-Do-Check-Act
PHD	Probabilistic Hypothesis Density
PJ	The Percentage of the Journey
PPG	Photoplethysmography
RBF	Radial Basis Function
RELU	Rectified Linear Unit Function
ROI	Region of Interest
ROS	Robot Operating Systems
RMSE	Root Mean Square Error

SDL	Scenario Definition Language
SDLP	The Standard Deviation of Lane Position
SLAM	Simultaneous Localization And Mapping
SORT	Simple Online and Real-time Tracking
SPSS	Statistical Product and Service Solutions
SRK	Skills, Rules, and Knowledge
TH	Time Headway
TOR	Take-over Request
TSR	Traffic Sign Recognition
TTC	Time To Collision
UKF	Unscented Kalman Filter
VRUs	Vulnerable Road Users

1

Continuous Game Theory Pedestrian Modelling Method for Autonomous Vehicles

Fanta Camara[1,2], Serhan Cosar[3], Nicola Bellotto[2], Natasha Merat[1] and Charles W. Fox[1,2,4]

[1]Institute for Transport Studies, University of Leeds, UK
[2]School of Computer Science, University of Lincoln, UK
[3]Institute of Engineering Sciences, De Montfort University, UK
[4]Ibex Automation Ltd., UK
E-mail: tsfc@leeds.ac.uk; serhan.cosar@dmu.ac.uk; nbellotto@lincoln.ac.uk; n.merat@its.leeds.ac.uk; chfox@lincoln.ac.uk

Autonomous Vehicles (AVs) must interact with other road users. They must understand and adapt to complex pedestrian behaviour, especially during crossings where priority is not clearly defined. This includes feedback effects such as modelling a pedestrian's likely behaviours resulting from changes in the AVs behaviour. For example, whether a pedestrian will yield if the AV accelerates, and vice versa. To enable such automated interactions, it is necessary for the AV to possess a statistical model of the pedestrian's responses to its own actions. A previous work demonstrated a proof-of-concept method to fit parameters to a simplified model based on data from a highly artificial discrete laboratory task with human subjects. The method was based on LIDAR-based person tracking, game theory, and Gaussian process analysis. The present study extends this method to enable analysis of more realistic *continuous* human experimental data. It shows for the first time how game-theoretic predictive parameters can be fit into pedestrians natural and continuous motion during road-crossings, and how predictions can be made about their interactions with AV controllers in similar real-world settings.

1.1 Introduction

Understanding pedestrian behaviour is now of upmost importance for Autonomous Vehicles (AVs) [5]. The potential future deployment of AVs is currently creating much enthusiasm [4, 43], as such vehicles would make transportation more efficient [22]. Huge improvements have been made on robotic localisation and mapping problems using simultaneous localisation and mapping (SLAM) algorithms [6, 38], together with new, cheap sensors, computation technologies, free and open-source software implementations [20, 42]. 'Self-driving' cars can now localise themselves and navigate by planning and controlling their routes on some roads, promising a future society with a better mobility system with less accidents and traffic in cities [22].

But before any fully self-driving revolution happens, AVs must share space with and will be challenged by human drivers and pedestrians, who are much harder to model and act upon than passive environments. Full self-driving must include this ability as well as the now-mature localisation, planning and routing technologies. Decades of research on human interaction in Transport Psychology and Human Factors has not yet been translated into robotic control systems, and many questions are still unanswered.

In most current 'self-driving' systems, for safety and legal reasons, pedestrians are considered as obstacles, such that the vehicle always stops for them. But recent real-world AV studies have shown that pedestrians may then take advantage of this predictable behaviour [27, 25, 5], pushing in front of them for priority eventually in *every* negotiation, so that the vehicles then make no progress. This has become known as the 'freezing robot problem (FRP)' [39].

Real human driving is massively more complex than simply mapping, localising and path planning. It is considered an art form by advanced practitioners such as members of the Institute for Advanced Motorists and other advanced drivers such as high-speed police and ambulance drivers [17]. In their training, these practitioners emphasise the human psychological processes involved in reading and predicting the behaviours of other road users as the most important skill of human drivers. Can you tell if a pedestrian is assertive enough to risk stepping out in front of you from their body language, their facial expressions, even their clothes and demographics? Road users have different utility functions, ranging from timid pedestrians likely to give way to all oncoming traffic, though to business-people late for a meeting or patients for an urgent medical appointment becoming much more assertive and risk-taking. Drivers must also consider the psychological effects of their

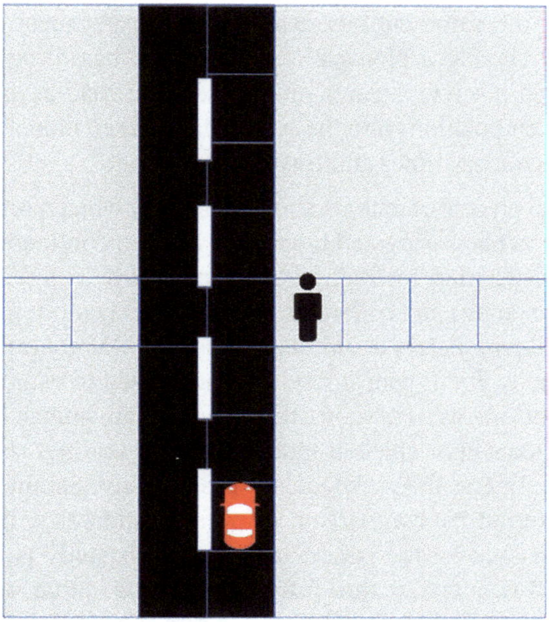

Figure 1.1 Two agents negotiating for priority at an intersection.

own actions. Speeding up and slowing down are not just ways to control one's own progress, but also send information about our own personality and risk preferences to pedestrians engaged in such negotiations for priority, along with other possible signals including lateral road positing, and more conventional signals such as flashing indicator lights and headlights, and driver face and arm expressions.

To progress towards automation of such understandings, Fox et al. [18] proposed and solved a simple game-theoretical mathematical model of the unsigned road-crossing scenarios represented in Figures 1.1 and 1.7. This model, based on the famous game of 'chicken', is called 'sequential chicken'. In this model, two agents – which may be pedestrians and/or vehicles – compete for space at an unsigned intersection, using only their positions to signal information to one another. Time, space and actions are discretised and it is assumed that both players have equal utility functions and know this to be the case. The model leaves open free parameters specifying the utility function for human players. Camara et al. [11] then asked human subjects to play sequential chicken as a board game, and developed a statistical method to fit parameters to the mathematical model to describe and predict their behaviours. In [9], the same authors extended this experiment to the case of human subjects playing a physical version of the board game, moving their bodies between discrete squares on and near the road at discrete time turns, integrating their positions into the sequential chicken model via LIDAR sensors, support vector machines and Bayesian tracking.

Contributions: The present chapter is a methods study which presents a new, full stack approach to measuring and modelling natural, continuous time and continuous space pedestrian interactions. It shows how to infer pedestrian preferences for time delays and collisions from their body motions as tracked by LIDAR. Inferred parameters could then be used in AV controllers during pedestrian interactions. First, pedestrian tracking is used to estimate the trajectories of the agents involved in semi-structured human–human interactions while playing the sequential chicken model. Second, optimal strategies are computed using the game theory model in [18]. Lastly, parameters of the interactions are inferred by comparison to optimal strategies, using Gaussian process regression over the parameter space. This study is intended to illustrate a proof-of-concept of this full-stack *method*: more detailed and controlled experiments will be needed to obtain robust parameters results and to learn about variations in parameters between different classes of pedestrians. The demonstrated method could also be used to model and measure pedestrian/pedestrian, human–driver–vehicle/AV, and human–driver/human–driver and AV/AV interactions as well as the primarily intended pedestrian/AV case.

This work is part of the EU H2020 interACT project with a consortium of European partners[1] investigating on the future deployment of AVs in mixed traffic environments with human drivers, cyclists and pedestrians. The overall aims of the project are to understand the behaviour of other road users, and how AVs could interact with them in a safe and efficient manner, and to propose new external Human–Machine Interface (eHMI) solutions that could facilitate the communication between AVs and people.

1.2 Related Work

1.2.1 Pedestrian Crossing Behaviour

A review on different approaches for pedestrian behaviour modelling is provided in [8]. Methods of pedestrian behaviour analysis are often performed

[1]https://www.interact-roadautomation.eu/

via video recording, semi-structured interviews and VR recording. Previous studies on pedestrian crossing behaviour can be found in [19, 29, 32]. For example, Gorrini et al. [19] analysed video data of interaction between pedestrians and vehicles at an unsignalized intersection using semi-automatic tracking. Their study showed that pedestrian crossing behaviour can be divided into three phases: approaching (stable speed), appraising (deceleration due to evaluation of speed and distance of oncoming vehicles) and crossing (acceleration). Papadimitriou et al. [29] compared observed and declared behaviour of pedestrians at different crossing areas, as a method to assess pedestrian risk-taking while crossing. They found that their observed behaviour is in accordance with their declared behaviours from a questionnaire survey and they report that female and male participants have similar crossing behaviour. Many studies such as [35] were focused on the evaluation of speed, TTC (Time To Collision), gap acceptance and communication means (e.g., eye contact and motion pattern) of the road users. Some other studies (e.g., [15]) have suggested that for autonomous vehicles, some apparently intuitive human communication styles might not be necessary for interactions with pedestrians. Dey and Terken [15] showed that facial communication cues such as eye contact do not play a major role in pedestrian crossing behaviour, and that the motion pattern and behaviour of vehicles are more important. The field study in [34] showed similar results with an 'unmanned' vehicle, suggesting that the same results could be found with autonomous vehicles. Risto et al. [33] showed that vehicle movement is sufficient for indicating the intention of drivers and presented some motion patterns of road users such as advancing, slowing early and stopping short.

1.2.2 Game Theory

Game theory offers a framework for modelling conflict and cooperation between rational decision-makers. It was developed in the 1940s by von Neumann and Morgenstern [28]. Its core concept is (Nash) *equilibrium* which is the pair of strategies (probability distributions over actions to be played) such that none of the players would change their strategy if they knew the other's strategy. Previous studies in Transport Studies and highway design have applied the game theory to several driver behaviour modelling tasks, as reviewed in [16]. Kim et al. [21] developed a mixed-motive game theory model for deciding the strategy chosen by two AVs equipped with adaptive cruise control (ACC). Meng et al. [26] also used the game theory for modelling AV lane-changing maneuvers. Rakha et al. [30] proposed a game theory approach for intersection conflicts management with reactive agents (the automated vehicles) equipped with ACC systems and a manager agent is used to decide the optimal strategy that increases the overall performance of all the agents. This approach prevents crashes from occurring and it also minimises the time delay in the intersection. Similar to our work, Ma et al. [24] computed Nash equilibria using Fictitious Play. Their method differs from ours in that not only their model takes into account pedestrians' position from a single image but also used some visual features from their appearance as part of the utility function to improve trajectory prediction. Adkins [1] presented an algorithm for intersection management involving up to four self-driving cars communicating with each other. Two motion choices are available for each player (move forward or stop) and an optimised solution using the game theory to solve the discrete intersection problem is presented. Turnwald et al. [40] proposed a non-cooperative game theoretic approach to human collision avoidance. Their method differs from ours in that they used a motion capture system to record human motions, a Bootstrap algorithm to compute the confidence intervals and applied a Dynamic Time Warping (DTW) algorithm to measure similarity between the trajectories. Variants of the game of chicken were proposed in [13, 27, 31] to solve conflicts between agents at intersections. A cellular automata approach was implemented in [31] and [13] for agents' interactions while [27] focused on the interaction between an AV and a pedestrian.

When multiple equilibria are present in games, standard game theory does not specify how the players should choose the best one. In the above studies, no method is proposed for the players to select which equilibrium to use. Typically this is because Transport Studies seeks to describe macroscopic flows of traffic rather than prescribe actions for individual vehicles, and considers that *any* possible equilibrium is a good description of observed data. For example in [27], the choice for the best solution depends on 'local social norms' which assumes that drivers should have prior knowledge of local customs. Unusually, [18] proposed a novel approach for optimal strategy prescription, called *meta-strategy convergence*. This method begins by choosing an equal-weighted mixture of strategies from all rational equilibria (after removing dominated and asymmetric equilibria where possible). The resulting strategies do not in general form an equilibrium themselves, but by applying fictitious play until convergence, a single equilibrium is obtained upon which it is argued that two rational players should agree without communication. Most of the game theory models reviewed earlier outperform non-game theoretic predictive models [13, 24, 30, 41].

1.2.3 Pedestrian Tracking

Pedestrian tracking plays an important role in many commercial applications but it is still a challenge for computer vision systems because of the multiple uncertainties (e.g., occlusions) due to complex environments [7]. Tracking of pedestrians requires the estimation of non-linear, non-Gaussian problems due to human motion, pedestrian scales and posture changes. Monte Carlo methods such as particle filtered-based approaches draw a set of samples assigned to a target and perform the data association for multiple targets using probabilistic techniques such as Nearest Neighbor (NN), Multiple Hypothesis Tracking (MHT), JPDAF and PHD-filter [2, 7]. Pedestrian tracking is composed of two steps: (i) a prediction step to determine the expected position and motion state and (ii) an update step to refine the prediction using sensor observations. Tracking has been previously combined with game theory for multi-robot system coordination problems. For instance, Skrzypczyk et al. [37] used non-cooperative games to control a team of mobile robots for a target tracking. When multiple equilibria are present, an arbiter based on the min-max method is used to fairly distribute costs among robots. Li et al. [23] applied cooperative game theory to improve tracking performance for a group of robots, allowing communication between the robots in order to minimise tracking costs and maximise the interests of the overall system of robots. Yan et al. [46] proposed a cooperative non-zero sum game approach for the problem of multi-target tracking for a multi-robot system in dynamic environment.

1.3 Methods

The present study demonstrates a method to fit parameters of the sequential chicken model to *continuous* human behaviour collected from controlled laboratory pedestrian–pedestrian interactions. The laboratory environment is designed to enable the simplest possible mapping of continuous physical human motions onto the model. Studying pedestrian–pedestrian interactions in place of pedestrian–AV interactions allows us to collect twice as much pedestrian data, and not require us to bias the experiment by involving an AV programmed with its own preferences.

1.3.1 Human Experiment

Eighteen human volunteer subjects (University of Lincoln Computer Science staff and students) were divided into nine pairs, one designated as player Y and the other as player X. Each pair was asked to play a physical version of

Figure 1.2 Two participants playing the game of chicken during the experiment.

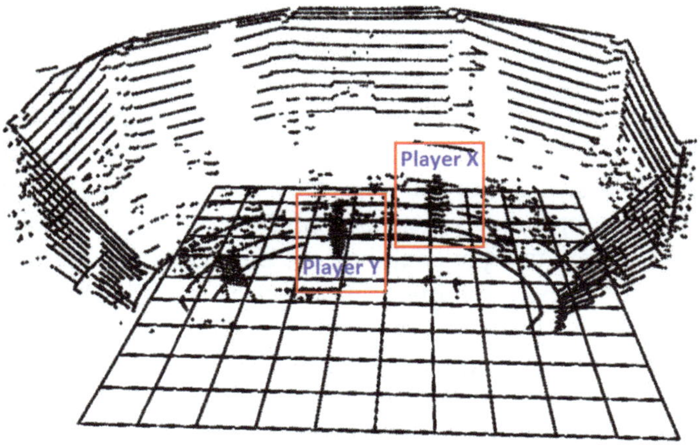

Figure 1.3 3D LIDAR output.

the sequential chicken game on a plus-maze shaped playing area drawn on an indoor floor as shown in Figure 1.2. Player Y was starting from $y = 6$ m and player X from $x = 6$ m such that they were both starting 6 m away from the intersection. Players were instructed that their objective was to pass the intersection as soon as possible, 'as if they were trying to reach their office entrance in a busy pedestrian area', on hearing the command 'go' to begin, given to both players at the same time. Each pair performed five interactions, i.e., 'games'. If both players walk at the same speed, then they collide with each other. Otherwise, one of them must yield to allow the other to pass the intersection point before them. Sometimes, both players try to yield at the same time, which does not break the symmetry, forcing them to continue negotiating one or more times. Players' motions were recorded using a Velodyne three-dimensional (3D) LIDAR. Figure 1.3 shows an example of the LIDAR output during the games.

1.3.2 Pedestrian Detection and Tracking

Pedestrian positions and velocities are provided by a robust Bayesian multi-target tracking systems based on 3D LIDAR detections [47], suitable for real-time, long-range tracking of multiple people in dynamic scenarios. Non-overlapping clusters of adjacent points are extracted based on their 3D Euclidean distance. An adaptive threshold accounts for the variation in shape and size of the human body in 3D LIDAR point clouds, which is a function of the person's distance from the sensor. Finally, clusters too large or too small to be humans are discarded by the detector, which outputs the distance and bearing of the cluster's centroid projected on the floor. The information from the detector is processed by a multi-target tracker, including an efficient implementation of Unscented Kalman Filter (UKF) and NN data association to deal with multiple detections simultaneously [3]. The tracker estimates the 2D coordinates and velocities of each pedestrian using a standard prediction-update recursive algorithm. The prediction step is based on the following constant velocity model,

$$\begin{cases} x_k = x_{k-1} + \Delta t\ \dot{x}_{k-1} \\ \dot{x}_k = \dot{x}_{k-1} \\ y_k = y_{k-1} + \Delta t\ \dot{y}_{k-1} \\ \dot{y}_k = \dot{y}_{k-1} \end{cases} \tag{1.1}$$

where x_k and y_k are the Cartesian coordinates of the target at time t_k, $\dot{x}_k$ and $\dot{y}_k$ are the respective velocities, and $\Delta t = t_k - t_{k-1}$. (The symbols x, y, t in this section are re–used to name different things than in the game theory model sections.) The update step of the estimation uses a 2D polar observation model to represent the position of a detected cluster,

$$\begin{cases} \phi_k & = \tan^{-1}(y_k/x_k) \\ \gamma_k & = \sqrt{x_k^2 + y_k^2} \end{cases} \tag{1.2}$$

where ϕ_k and γ_k are, respectively, the bearing and the distance of the cluster's centroid with respect to the sensor. More details can be found in [3, 47].

Figures 1.4 to 1.6 show the filtering process for pedestrian tracks. Like all detection and tracking methods, the system sometimes produces false positives and false negatives. To remove false positives, tracks were filtered to exclude those including any locations outside the plus-maze area, as shown in

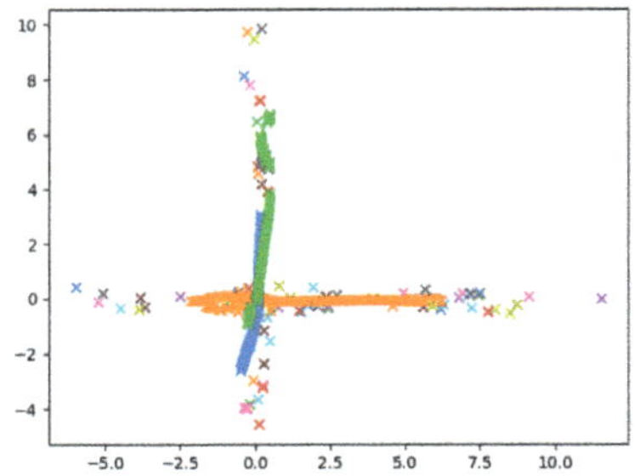

Figure 1.4 Unfiltered tracks.

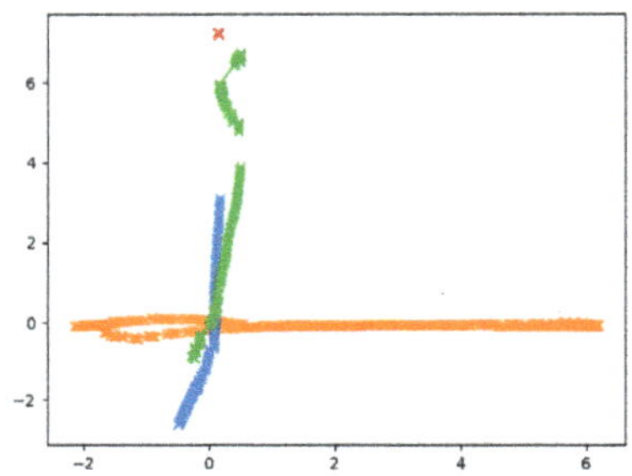

Figure 1.5 Filtered tracks.

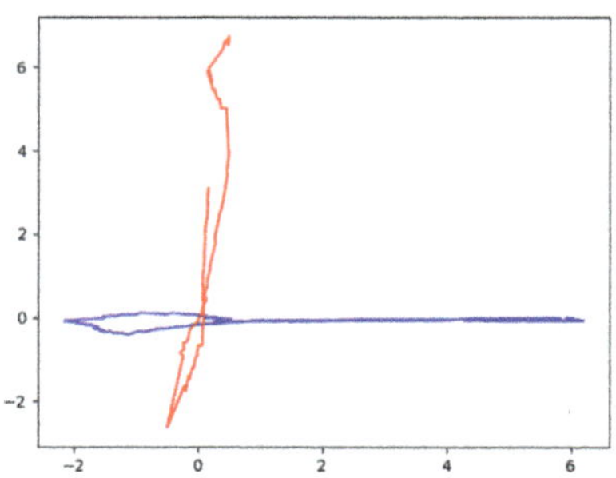

Figure 1.6 Tracks assigned to players.

Figure 1.5. Due to occasional false positives with tracks, and false negatives missing tracks, filtering resulted in a collection of 14 games, from 6 different pairs of players, having good and complete tracks for both players together, that are used in the rest of the analysis.

1.3.3 Sequential Chicken Model

In sequential chicken, two agents called Y and X are driving straight towards each other at right angles as in Figure 1.1, such that they will collide unless one of them yields to the other. The sequential chicken model operates on discrete *space* as in Figure 1.7; discrete *times* ('turns') during which the agents can adjust their discrete *speeds*, simultaneously selecting between speeds of either 1 square per turn or 2 squares per turn, at each turn. Both agents want to pass the intersection as soon as possible to avoid travel delays, but if they collide, they are both bigger losers as they both receive a negative utility U_{crash}. Otherwise if the players pass the intersection, each receives a time delay penalty $-TU_{\text{time}}$, where T is the time from the start of the game and U_{time} represents the value of saving one turn of travel time. The model assumes that the two players choose their actions (speeds) $a_Y, a_X \in \{1, 2\}$ simultaneously, then implement them simultaneously, at each of several discrete-time turns. There is no lateral motion (positioning within the lanes of the roads) or communication between the agents other than via their visible positions. The game is symmetric, as both players are assumed to know that they have the same utility functions ($U_{\text{crash}}, U_{\text{time}}$), hence they both have the same optimal strategies. These optimal strategies are derivable from the game theory together with meta-strategy convergence, via recursion [18]. Sequential chicken can be viewed as a sequence of one-shot sub-games, whose payoffs are the expected values of new games resulting from the actions, and are solvable by standard game theory.

Discretised locations of the players can be represented by (y, x, t) at discretised turn t and their discretised actions $a_Y, a_X \in \{1, 2\}$ for speed selection. Similar to the approach used in [10], discretisations are obtained from the continuous data by quantizing continuous position into about 0.1 m

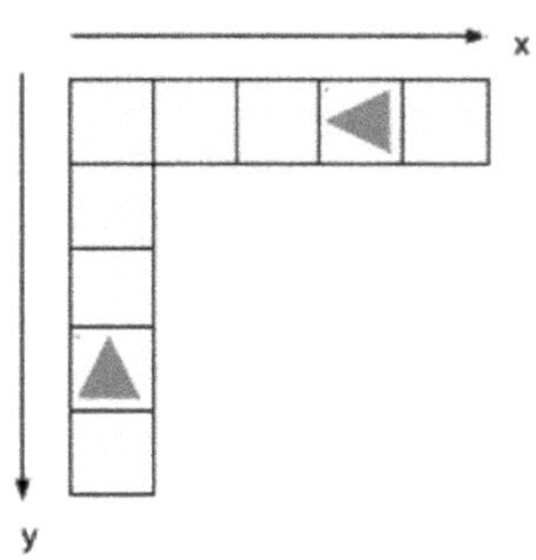

Figure 1.7 Sequential chicken game.

locations every 0.09 s turn, by averaging over all locations during that interval; and quantizing actions into SLOW or FAST between each pair of quantised locations according to whether the location change is greater or lower than a 1 m/s threshold.

The new state at turn $t+1$ is given by $(y + a_Y, x + a_X, t+1)$. Define $v_{y,x,t} = (v^Y_{y,x,t}, v^X_{y,x,t})$ as the value (expected utility, assuming all players play optimally) of the game for state (y, x, t). As in the standard game theory the value of each 2×2 payoff matrix can then be written as,

$$v_{y,x,t} = v\left(\begin{bmatrix} v(y-1, x-1, t+1) & v(y-1, x-2, t+1) \\ v(y-2, x-1, t+1) & v(y-2, x-2, t+1) \end{bmatrix}\right), \tag{1.3}$$

which can be solved using dynamic programming assuming meta-strategy convergence equilibrium selection. Under some approximations based on the temporal gauge invariance described in [18], we may remove the dependencies on the time t in our implementation so that only the locations (y, x) are required in computation of $v_{y,x}$ and optimal strategy selection.

In the sequential chicken model, if the two players play optimally, then there must exist a non-zero probability for a collision to occur. Intuitively, if we consider an AV to be one player that always yields, it will make no progress as the other player will always take advantage over it, hence there must be some threats of collision [18].

1.3.4 Gaussian Process Parameter Posterior Analysis

We use Gaussian processes regression [45] to fit the posterior belief over the behavioural parameters of interest, $\theta = (U_{crash}, U_{time})$ from the observed data, D. Under the sequential chicken model, M, these are

$$P(\theta|M, D) = \frac{P(D|\theta, M)P(\theta|M)}{\sum_{\theta'} P(D|\theta', M)P(\theta'|M)}. \tag{1.4}$$

We assume a flat prior over θ so that,

$$P(\theta|M, D) \propto P(D|\theta, M), \tag{1.5}$$

which is the data likelihood, given by,

$$P(D|\theta, M) = \prod_{\text{game}} \prod_{\text{turn}} P(d_Y^{\text{game,turn}}|y, x, \theta, M')P(d_X^{\text{game,turn}}|y, x, \theta, M'), \tag{1.6}$$

where $d_{\text{player}}^{\text{game,turn}}$ are the observed action choices, and y and x are the observed player locations at each $turn$ of each $game$. Here M' is a noisy version of the optimal sequential chicken model M, which plays actions from M with probability $(1 - s)$ and maximum entropy random actions (0.5 probability of each speed) with probability s. This modification is necessary to allow the model to fit data where human players have made deviations from optimal strategies which would otherwise occur in the data with probability zero. Real humans are unlikely to be perfectly optimal at anytime as they may make mistakes of perception and decision-making. This is a common method to weaken psychological models to allow non-zero probabilities for such mistakes if present.

For a given value of θ, we may compute the optimal strategy for the game by dynamic programming as in Algorithm 1. Optimal strategies are in general probabilistic, and prescribe the $P(d_Y^{\text{game,turn}}|y,\ x, \theta, M)$, $P(d_X^{\text{game,turn}}|y, x, \theta, M)$ terms to compute the above data likelihood. We then use a Gaussian process with a Radial Basis Function (RBF) kernel to smooth the likelihood function over all values of θ beyond a sample whose values are computed explicitly. In practice, this is performed in the log domain to avoid numerical computation problems with small probabilities. The resulting Gaussian process is then read as the (un-normalized, log) posterior belief over the behavioural parameters $\theta = \{U_{\text{time}}, U_{\text{crash}}\}$ of interest.

Algorithm 1 Optimal solution computation

for U_{crash} in range($U_{crash_{min}}$, $U_{crash_{max}}$) **do**
  for U_{time} in range($U_{time_{min}}$, $U_{time_{max}}$) **do**
    S ← strategy matrix($NY \times NX \times 2$) for P(player X chooses speed $2|y, x$)
    loglik = 0
    for each game in data **do**
      for each turn in game **do**
        loglik = $\prod_{game} \prod_{turn} (1-s)P(d_Y^{game,turn}|y,x,\theta,M)P(d_X^{game,turn}|y,x,\theta,M) + s(\frac{1}{2})$
      end for
    end for
    Store loglik(U_{crash}, U_{time})
  end for
end for
maxloglik ← max of loglik(U_{crash}, U_{time})

1.4 Results

After applying Gaussian process regression and optimising s to maximise the likelihood at the Maximum A Posteriori (MAP) point of θ, the posterior

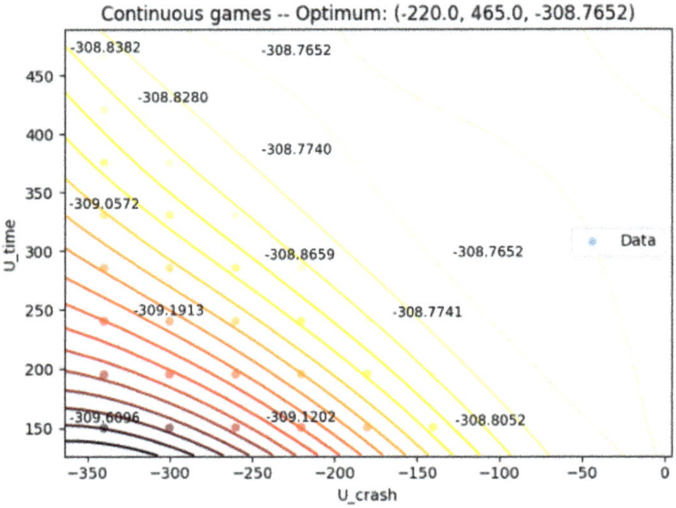

Figure 1.8 Gaussian process log-posterior over behavioural parameters.

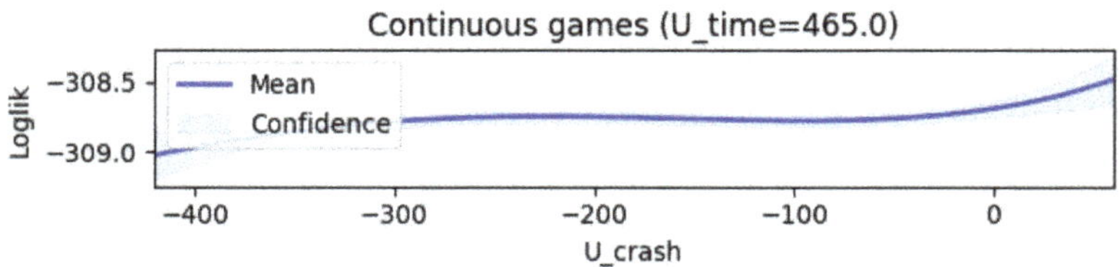

Figure 1.9 Slice through the Gaussian process showing standard deviation log-posterior confidence.

distribution over $\theta = \{U_{\text{crash}}, U_{\text{time}}\}$ is shown in Figure 1.8. The MAP estimate of the parameters is then around $U_{\text{crash}} = -220$, $U_{\text{time}} = 465$, at $s = 0.11$. The $-44 : 93 \simeq -1 : 2$ ratio in the utilities means that assuming the noisy model M' the subjects value about a 1/2 turn time delay equally to a crash, and the s value means that the subjects make mistakes from optimal behaviour in 11% of actions. Significance of the results can be seen by inspection of the thin standard deviation widths of 1D slices through the 2D posterior as in Figure 1.9. We can only see a small deviation when U_{crash} is too small or too large.

The behavioural parameter ($\theta \simeq -\frac{1}{2}$) shows that participants were having higher preferences for time saving rather than for collision avoidance, which is similar to the findings in [9, 11]. As in these studies, the high ratio may be explained by the artificial laboratory nature of the environment: subjects want to win the game and know there is no significant negative utility for collisions as the laboratory environment is designed to be safe. The method is now well developed enough to move to the real world for future studies, and we expect to see lower ratios there, where the cost of collisions with vehicles and other pedestrians is much higher.

1.5 Discussion

The results shown are from a small sample of data and are intended as a proof-of-concept of the proposed method. This shows how a full stack of real-time detection and tracking, and game theoretic modelling can work together to understand and predict continuous pedestrian interactions with another road user. The data used here is from pedestrian–pedestrian interactions and is only from a small sample of 14 interactions. Previous work performed this on highly artificial discrete time, turn taking human experiments. This is the first time that a method now exists for more natural continuous data as would be found in real-world AV interactions. The key concept in moving from discrete to continuous data is that we were able to discretise both players actions into just two discrete categories, SLOW and FAST, which enables the sequential chicken model to then operate with minimal changes.

Future work could now make use of this method, firstly to collect and analysis much larger experimental pedestrian–pedestrian data sets; and secondly to deploy a model trailed from this data as a controller in a real AV. It is possible that when trained on larger data sets, the model might show different preferences for different types of pedestrians. For example, real-time detectable features such as age [36], gender [48], body pose [12], activity recognition [14], gait [44] and style of dress might give information about pedestrian intention and behavioural preferences, which if found from training data could then be used to refine real-time AVs pedestrian predictions and active speed controls. This method could then possibly enable new AV online-learning algorithms that adapt to the environment or passenger's preferences.

Acknowlegdment

This project has received funding from the EU H2020 interACT project: Designing cooperative interaction of automated vehicles with other road users in mixed traffic environments under grant agreement No. 723395.

References

[1] R. Adkins. Autonomous vehicle intersection management. Technical report, 2016.

[2] N. Bellotto, S. Cosar, and Z. Yan. *Human Detection and Tracking*, pages 1–10. Springer Berlin Heidelberg, Berlin, Heidelberg, 2018.

[3] N. Bellotto and H. Hu. Computationally efficient solutions for tracking people with a mobile robot: an experimental evaluation of Bayesian filters. *Autonomous Robots*, 28:425–438, 2010.

[4] B. Bontrager. The race to fully autonomous cars, 2018. https://medium.com/swlh/the-race-to-fully-autonomous-cars-8212ff73aad.

[5] R. Brooks. The big problem with self-driving cars is people. In *IEEE Sprectrum*, 2017.

[6] C. Cadena, L. Carlone, H. Carrillo, Y. Latif, D. Scaramuzza, J. Neira, I. Reid, and J. Leonard. Past, present, and future of simultaneous localization and mapping: Towards the robust-perception age. *IEEE Transactions on Robotics* , 2016.

[7] F. Camara, N. Bellotto, S. Cosar, D. Nathanael, M. Althoff, J. Wu, J. Ruenz, A. Dietrich, and C. W. Fox. Pedestrian models for autonomous driving Part I: low-level models, from sensing to tracking. *IEEE Transactions on Intelligent Transportation Systems*, 2020. DOI: 10.1109/TITS.2020.3006768

[8] F. Camara, N. Bellotto, S. Cosar, F. Weber, D. Nathanael, M. Althoff, J. Wu, J. Ruenz, A. Dietrich, A. Schieben, G. Markkula, F. Tango, N. Merat, and C. W. Fox. Pedestrian models for autonomous driving Part II: high-level models of human behavior. *IEEE Transactions on Intelligent Transportation Systems*, 2020. DOI: 10.1109/TITS.2020.3006767

[9] F. Camara, S. Cosar, N. Bellotto, N. Merat, and C. W. Fox. Towards pedestrian-AV interaction: method for elucidating pedestrian preferences. In *IEEE/RSJ Intelligent Robots and Systems (IROS) Workshops*, 2018.

[10] F. Camara, N. Merat, and C. W. Fox. A heuristic model for pedestrian intention estimation. In *IEEE 22nd Intelligent Transportation Systems Conference (ITSC)*, 2019.

[11] F. Camara, R. Romano, G. Markkula, R. Madigan, N. Merat, and C. W. Fox. Empirical game theory of pedestrian interaction for autonomous vehicles. In *Measuring Behavior 2018: 11th International Conference on Methods and Techniques in Behavioral Research*, March 2018.

[12] Z. Cao, T. Simon, S.-E. Wei, and Y. Sheikh, “OpenPose: Realtime multi-person 2D pose estimation using part affinity fields,” in Proc. IEEE Conf. Comput. Vis. Pattern Recognit. (CVPR), 2017, pp. 7291–7299.

[13] P. Chen, C. Wu, and S. Zhu. Interaction between vehicles and pedestrians at uncontrolled mid-block crosswalks. *Safety Science*, 82:68 – 76, 2016.

[14] C. Coppola, D. R. Faria, U. Nunes, and N. Bellotto. Social activity recognition based on probabilistic merging of skeleton features with proximity priors from RGB-D data. In *2016 IEEE/RSJ International Conference on Intelligent Robots and Systems (IROS)*, pages 5055–5061, Oct 2016.

[15] D. Dey and J. Terken. Pedestrian interaction with vehicles: Roles of explicit and implicit communication. In *Proceedings of the 9th International Conference on Automotive User Interfaces and Interactive Vehicular Applications*, AutomotiveUI ’17, pages 109–113, New York, NY, USA, 2017. ACM.

[16] R. Elvik. A review of game-theoretic models of road user behaviour. *Accident Analysis & Prevention*, 62:388 – 396, 2014.

[17] Institute for Advanced Motorists. *How to Be A Better Driver: Advanced Driving the Essential Guide*. Motorbooks, 2007.

[18] C. W. Fox, F. Camara, G. Markkula, R. Romano, R. Madigan, and N. Merat. When should the chicken cross the road?: Game theory for autonomous vehicle - human interactions. In *Proceedings of VEHITS 2018: 4th International Conference on Vehicle Technology and Intelligent Transport Systems*, January 2018.

[19] A. Gorrini, G. Vizzari, and S. Bandini. Towards modelling pedestrian-vehicle interactions: Empirical study on urban unsignalized intersection. *CoRR*, abs/1610.07892, 2016.

[20] S. Kato, E. Takeuchi, Y. Ishiguro, Y. Ninomiya, K. Takeda, and T. Hamada. An open approach to autonomous vehicles. *IEEE Micro*, 35(6):60–68, Nov 2015.

[21] R. Kim, Changwon & Langari. Game theory based autonomous vehicles operation. *International Journal of Vehicle Design (IJVD), Vol. 65*, 2014.

[22] T. J. Kim. Automated autonomous vehicles: Prospects and impacts on society. *Journal of Transportation Technologies*, 2018.

[23] Y. Li, M. Li, L. Dou, Q. Zhao, Z. Wang, and J. Li. A socially multi-robot target tracking method based on extended cooperative game theory.

In *2014 IEEE International Conference on Robotics and Biomimetics (ROBIO 2014)*, pages 247–252, Dec 2014.
[24] Ma, Wei-Chiu, De-An Huang, Namhoon Lee, and Kris M. Kitani. "Forecasting interactive dynamics of pedestrians with fictitious play." In Proceedings of the IEEE Conference on Computer Vision and Pattern Recognition, pp. 774–782. 2017.
[25] R. Madigan, T. Louw, M. Dziennus, T. Graindorge, E. Ortega, M. Graindorge, and N. Merat. Acceptance of automated road transport systems (ARTS): An adaptation of the utaut model. *Transportation Research Procedia*, 14:2217 – 2226, 2016. Transport Research Arena TRA2016.
[26] F. Meng, J. Su, C. Liu, and W. H. Chen. Dynamic decision making in lane change: Game theory with receding horizon. In *2016 UKACC 11th International Conference on Control (CONTROL)*, pages 1–6, Aug 2016.
[27] A. Millard-Ball. Pedestrians, autonomous vehicles, and cities. *Journal of Planning Education and Research*, 38(1):6–12, 2018.
[28] O. Morgenstern and J. Von Neumann. *Theory of games and economic behavior*. Princeton university press, 1953.
[29] E. Papadimitriou, S. Lassarre, and G. Yannis. Pedestrian risk taking while road crossing: A comparison of observed and declared behaviour. *Transportation Research Procedia*, 14:4354 – 4363, 2016. Transport Research Arena TRA2016.
[30] H. Rakha, I. H. Zohdy, and R. K. Kamalanathsharma, "Agent-based gametheory modeling for driverless vehicles at intersections," Mid-Atlantic Univ. Transp. Center, Philadelphia, PA, USA, Tech. Rep. VT-2010-02, Feb. 2013.
[31] A. Rane, S. Krishnan, and S. Waman. Conflict resolution of autonomous cars using game theory and cellular automata. In *2014 International Conference on Reliability Optimization and Information Technology (ICROIT)*, pages 326–330, Feb 2014.
[32] A. Rasouli, I. Kotseruba, and J. K. Tsotsos. Agreeing to cross: How drivers and pedestrians communicate. *2017 IEEE Intelligent Vehicles Symposium (IV)*, pages 264–269, 2017.
[33] M. Risto, C. Emmenegger, E. Vinkhuyzen, M. Cefkin, and J. Hollan. Human-vehicle interfaces: The power of vehicle movement gestures in human road user coordination. pages 186–192, 11 2017.
[34] D. Rothenbücher, J. Li, D. Sirkin, B. Mok, and W. Ju. Ghost driver: A field study investigating the interaction between pedestrians and driverless vehicles. In *2016 25th IEEE International Symposium on Robot and*

Human Interactive Communication (RO-MAN), pages 795–802, Aug 2016.

[35] F. Schneemann and I. Gohl. Analyzing driver-pedestrian interaction at crosswalks: A contribution to autonomous driving in urban environments. In *2016 IEEE Intelligent Vehicles Symposium (IV)*, pages 38–43, June 2016.

[36] S. Sithungu and D. Van der Haar. Real-time age detection using a convolutional neural network. In W. Abramowicz and R. Corchuelo, editors, *Business Information Systems*, pages 245–256, Cham, 2019. Springer International Publishing.

[37] K. Skrzypczyk. Game theory based target following by a team of robots. In *Proceedings of the Fourth International Workshop on Robot Motion and Control (IEEE Cat. No.04EX891)*, pages 91–96, June 2004.

[38] S. Thrun, W. Burgard, and D. Fox. *Probabilistic Robotics*. Intelligent robotics and autonomous agents. MIT Press, 2005.

[39] P. Trautman and A. Krause. Unfreezing the robot: Navigation in dense, interacting crowds. In *2010 IEEE/RSJ International Conference on Intelligent Robots and Systems*, pages 797–803, Oct 2010.

[40] A. Turnwald, D. Althoff, D. Wollherr, and M. Buss. Understanding human avoidance behavior: Interaction-aware decision making based on game theory. *International Journal of Social Robotics*, 8(2):331–351, Apr 2016.

[41] A. Turnwald, W. Olszowy, D. Wollherr, and M. Buss. Interactive navigation of humans from a game theoretic perspective. In *2014 IEEE/RSJ International Conference on Intelligent Robots and Systems*, pages 703–708, Sept 2014.

[42] Udacity. Open source self-driving car, 2017.

[43] M. Wade. Silicon valley is winning the race to build the first driverless cars, February 2018. theconversation.com.

[44] L. Wang, T. Tan, H. Ning, and W. Hu. Silhouette analysis-based gait recognition for human identification. *IEEE Transactions on Pattern Analysis and Machine Intelligence*, 25(12):1505–1518, Dec 2003.

[45] C. K. Williams and C. E. Rasmussen. Gaussian processes for regression. In *Advances in neural information processing systems*, pages 514–520, 1996.

[46] M. Yan. Multi-robot searching using game-theory based approach. *International Journal of Advanced Robotic Systems*, 5(4):44, 2008.

[47] Z. Yan, T. Duckett, and N. Bellotto. Online learning for human classification in 3d LIDAR-based tracking. In *IEEE/RSJ Int. Conf. on Intelligent Robots and Systems (IROS)*, pages 864–871, 2017.

[48] W. Zhang, M. L. Smith, L. N. Smith, and A. Farooq. Gender recognition from facial images: two or three dimensions? *JOSA A*, 33(3):333–344, 2016.

2

The Interface Challenge for Partially Automated Vehicles: How Driver Characteristics Affect Information Usage Over Time

Arun Ulahannan[1], Paul Jennings[2], Simon Thompson[3] and Stewart Birrell[4]

[1]Coventry University, United Kingdom
[2]WMG, University of Warwick, United Kingdom
[3]Jaguar Land Rover, United Kingdom
[4]Coventry University, United Kingdom
E-mail: Arun.Ulahannan@coventry.ac.uk; Paul.Jennings@warwick.ac.uk; S.Birrell@coventry.ac.uk; sthom261@jaguarlandrover.com
Research supported by Jaguar Land Rover (JLR), UK

Understanding how best to present information inside a partially automated vehicle is a prevalent challenge in Human–Machine Interface (HMI) design. To date little is known about how characteristics around trust, driving experience and cognitive workload specifically affect the types of information that should be presented in an automated vehicle. It is also unknown how these requirements change with increasing familiarity with the system.This two-part driving simulator study aimed to understand how trust and perceived workload changed with increasing exposure to a partially automated vehicle and how this corelated with information usage. Forty-four participants experienced nine partially automated simulated driving scenarios over the course of three or five consecutive sessions across the two studies. Eye tracking was used to record the information observed. Participants were asked to complete the Jian Trust Questionnaire, Driver Behaviour Questionnaire (DBQ) and the Driver Activity Load Index (DALI). Significant changes to trust and perceived workload were observed. Workload was found to decrease with lower fixations to information around the monitoring task.Drivers who were more prone to lapses or errors (as measured by the DBQ) tended towards less cognitively demanding information (skill based). This study has contributed to a better understanding of how driver characteristics can affect information use inside partially automated vehicles and such factors must be considered in future HMI design.

2.1 Introduction

Partially automated vehicles (NHTSA Levels 2–4) are increasing in prevalence both in research and in commercial availability [1]. In such vehicles, particularly at the lower levels of automation, passengers will be required to monitor the automated process in the event that control must be handed back to the human driver. There has been increasing evidence to suggest that humans are not efficient at monitoring automated processes [2–4] and this can promote riskier driving behaviour inside vehicles [5].

Appropriately designed Human–Machine Interfaces (HMI)have been shown to play an important role in mitigating these challenges drivers could face in partially automated systems [6, 7]. One of the underlying reasons for this is trust and how successfully the HMI can support the driver in helping them place the right level of trust in the vehicle's capabilities [8].

In partially automated vehicles today, HMI designers have attempted to address the problem by providing the driver with many different types of information [9], however, there is a consensus that this approach can cause cognitive overload and create the conditions for vehicle accidents [10–12].

Further, different drivers place different levels of trust in partially automated vehicles and have different driving styles. The process of trust development is also dynamic and changes over time [13]. Furthermore,it has been found that information requirements do vary between drivers [14] and these requirements change over time as the driver becomes more accustomed to the system [15]. There are models that recognize the differences in cognitive processing between different drivers. Models such as Skills, Rules, Knowledge (SRK) [16]; Strategic, Manoeuvring, Control [17] and Primary, Secondary, Tertiary [18] all attempt to categorize the cognitive processes that occur in the user in response to different types of information. There would appear to be an opportunity to understand how the information presented to the driver can be adapted over time to match driver preferences. Having the right information presented at the appropriate time could help build and maintain trust in the vehicle, which consequently promotes the driver's 'appropriate use' of partially automated vehicles. There appeared to be a dearth of literature that has considered the idea of an adaptive interface, with some preliminary studies supporting the idea that it helped reduce the driver's workload [12, 19]. In a review into the adaptive interfaces for complex systems, it was concluded that more research was required into user–centric design in adaptive systems [20].

This study aims to understand how trust and perceived workload and a driver's driving behaviour relate to the information they require inside a partially automated vehicle over multiple exposures to the partially automated system. This makes this study unique in that it evaluates changes in information requirements over time and not just from a single exposure to an automated vehicle simulator trial.

2.2 Method

To understand how information requirements are affected by trust, perceived workload and a driver's driving behaviour, a two part study design was used.

2.2.1 Study Design

Both Parts 1 and 2 used a longitudinal within-subjects design over either five (Part 1) or three days (Part 2). A summary of the two study parts can be seen in Table 2.1. The participants were placed into a driving simulator and presented with two to three, 5–13 minutes partially automated driving scenarios. At the same time, an interface with nine pieces of information representing various aspects of the vehicle's sensors and intentions was displayed (Table 2.3).

In Part 1, the participants were given the Jian Trust Questionnaire and DBQ (Driver Behaviour Questionnaire). In Part 2, the participants were given the Jian Trust Questionnaire and DALI (Driver Activity Load Index) questionnaires. A more detailed summary of each questionnaire can be found in Section 2.2.5. A summary of the study design can be seen in Table 2.1.

It was found in the first study (5–day design) that fixations exhibited most change on days 1, 3 and 5. Furthermore, because the number of simulations remained consistent (Nine simulations per participant), the 3_day design in Part 2 enabled the flexibility to ensure more participants were able to complete all nine simulations.

2.2.2 Participants

A total of 44 participants took part in the study, recruited through advertising (using social media and flyers) at the Warwick University campus and the local area. A detailed breakdown of the participants can be seen in Table 2.2.

Table 2.1 Summary of study design

	Session	
Study	Part 1	Part 2
Date	July 2018	March 2019
Number of Participants	17	27
Simulation	9 scenarios over 5 days	9 scenarios over 3 days
Information	9 information types	9 information types
Measurement	SMI Glasses	Tobii Pro 2
Scenario	Steady State Partially Automated	Steady State Partially Automated
Questionnaires	Jian Trust&Driver Behaviour Questionnaire	Jian Trust&Driver Activity Load Index

Table 2.2 Breakdown of participant demographics

Information	Part 1	Part 2
Gender	8 (Male), 9 (Female)	14 (Male), 13 (Female)
Age	2 (18-24), 11 (25-34), 1 (66-64), 3 (65 or older)	10 (18-24), 13 (25-34), 1 (41-50), 3 (71-80)

During Part 1, three participants had to withdraw from the study, as a result, these participants were omitted from the results. All participants in Part 2 were able to complete all their sessions.

2.2.3 Interface Design

Across both Parts 1 and 2 of the study, the same interface and information was presented to the participants.

First the information that would be displayed had to be chosen. Standards such as BS EN ISO 15008:2017 [21] and UNECE 121 [22] were reviewed. These define the minimal information requirements for vehicles today and other factors such as legibility, image quality, etc. However, to narrow down the amount of information presented, they were then categorized against Rasmussen's Skills, Rules, Knowledge (SRK) model [16]. The SRK model proposed that information is acted on by users in different ways. In this paper, this model was used to categorize the information into its expected cognitive workload. Skill–based information is acted on intuitively and requires little cognitive processing from the driver. Rule–based information is slightly more complex, requiring the driver to understand the information and recall on previous behaviour or experience. Knowledge–based information is the

Table 2.3 Information categorised by SRK

Skills	Rules	Knowledge
Automated Driving Indicator	Battery Charge Level	Hazard Sensor Display
Road Signs	Action Explanation	Vehicle Warnings
Traffic Conditions	Navigation	Energy Usage

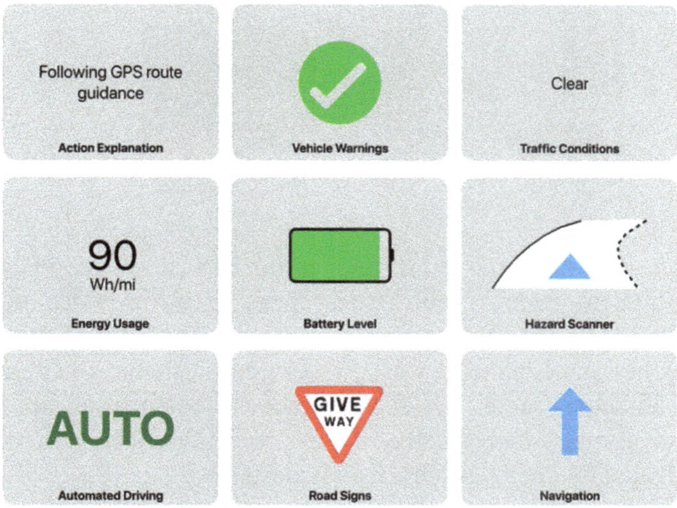

Figure 2.1 Final interface presented to the participants.

most cognitively demanding, requiring the driver to interpret the information, develop a mental model of what is presented and then relate the information to the real world conditions.

Using the SRK [16] and the methodologies used in [15], a methodologically derived shortlist of information types was developed. The final categorizations were then decided upon by a group of academics from WMG, University of Warwick and industry HMI experts. The final categorizations are shown in Table 2.3.

The interface displayed (forday one, session one) to the participants can be seen in Figure 2.1.

The interface was then displayed on a 10.5" iPad Pro. The information displayed on the screen would update in real time, according to the conditions in the simulation.

The interface was prototyped using tachistoscopic presentation to understand if any of the icons were visually more salient than others. This led

Figure 2.2 WMG 3xD development simulator.

to the redesign of some of the icons- namely the hazard sensor, which was changed from a photo–realistic vehicle to a generic red arrow. Visual salience is primarily guided by the relative similarity (or dissimilarity) of the icons' attributes [23], hence the interface was also designed such that all animations animated at the same frames per second value to balance the animation salience. It was also considered that any effects of imbalanced icon salience would be mitigated by the longitudinal study design.

2.2.4 Driving Scenario

The driving scenario in both Parts 1 and 2 of the study was presented on the WMG Development simulator and using XPI Simulation software. This was a three screen immersive setup. This study only focused on the steady–state portions of both Parts 1 and 2. The scenarios varied between rural, urban and motor way environments. The simulation setup can be seen in Figure 2.2.

2.2.5 Procedure

Procedurally, both Parts 1 and 2 were the same with the only differences being the number of trial days and the driving scenarios presented. In this study, a total of three questionnaires were presented to the participants for completion:

- The Jian Trust Questionnaire [24] is an established method of determining the trust a participant has in an automated system, measured on a seven–point Likert scale. This was completed three times in both Parts 1 and 2 of the study at the end of the simulator sessions. As a result of Part

1's 5–day design, the questionnaire was given on alternate sessions (1, 3 and 5). This was to avoid questionnaire fatigue. Part 2's 3–day design meant the questionnaire was given on all days. The result was a set of three results per study.

- The DBQ [25] classifies different driving behaviours into errors, lapses and violations. Errors are defined as being unintentional but having potentially dangerous consequences. Lapses are unintentional and of no serious consequence whereas violations are intentional actions that are dangerous. Each of these is measured on a five-point Likert scale. The DBQ was completed only in Part 1 of the study and only once in the trial week before beginning the simulation. This was because the DBQ is concerned with the participants' past driving experience and this will not have changed over the course of the five study days.
- The DALI [26] is a measure of perceived workload that has been adapted from the NASA TLX questionnaire, but is more suited for the context of vehicle workload. The DALI was given to the participants only during Part 2 and at the end of every simulation session (hence three times in the trial week).

Participants were brought into the simulator room and given a briefing on the experimental procedure. They were told that the simulator was a partially automated vehicle and were given an explanation of the capabilities of the vehicle. They were told to observe the simulated scenario as if they were the driver of the vehicle in partially automated driving mode and to use the information presented to them in any way that made them feel comfortable in the vehicle.

Participants were fitted with wearable eye tracking glasses. In Part 1, SMI glasses were used which recorded at 30 Hz. In Part 2, the Tobii Pro 2 glasses were used and recorded at 100 Hz. In both cases,the glasses meant that the participants were not obliged to keep their head in a particular reference frame and were free to move their head. For this paper's analysis, fixations of 200 ms or longer were used. Fixations below 200 ms are not long enough to assume cognitive processing of the information [27–29], hence this was used as the minimum fixation threshold.

Part 1 investigated how trust and driver behaviour could affect information requirements. Part 2 also investigated information requirements but looked at the effect of trust and perceived workload using the DALI.Across both Parts 1 and 2, the scheduled meeting time for the daily sessions were kept consistent for each participant.

2.2.6 Data Analysis

The aim of this study was to understand how the driver's trust, perceived workload and driving behaviour relate to their information requirements inside a partially automated vehicle. For comparisons within the measures of trust, DBQ and DALI, either the Repeated Measures ANOVA or the non-parametric Wilcoxon Test was used depending on whether the assumption of normality was violated by the data. For the analysis of the relationship between measures, the Pearson or non-parametric Spearman correlation tests were used accordingly.

2.2.6.1 Trust (Parts 1 and 2)

Trust was captured three times during the trial week. To simplify analysis, distrust scores were reversed and converted to a trust score (by taking away the result from seven). These scores were then averaged to create a combined trust score and the results are presented in Table 2.4 (where 1 is very low trust and 7 is very high trust). Trust was then correlated against the fixations, DBQ and DALI.

2.2.6.2 DBQ (Part 1 only)

The DBQ was split into errors, lapses and violations. The results are listed in Table 2.5. This result was then correlated against fixations and trust but not the DALI (as DALI was only presented in Part 2). However, because the DBQ was only captured once at the start of the trial week, the results were correlated against the total number of fixations for the week for each participant to each piece of information.

2.2.6.3 DALI (Part 2 only)

The DALI measured perceived workload based on six factors (effort of attention, visual demand, auditory demand, temporal demand, interference and situations stress) [26] (where 0 is no cognitive workload and 100 is maximum cognitive workload). The DALI was captured three times during the trial week. Table 2.6 shows the DALI results for the trial week. This was then correlated against fixations and trust but not the DBQ.

2.3 Results

In the following tables, all significant results are denoted with an asterix (*).

Table 2.4 Trust results from Parts 1 and 2

	Measure		
Trust Results	1	2	3
N	44	44	44
Mean	4.54	4.81	4.87
Std. Dev.	1.02	1.02	1.14

Table 2.5 DBQ results from Part 1

	Measures		
DBQ Results	Lapses	Violations	Errors
N	17	17	17
Mean	1.43	0.91	0.61
Std. Dev	0.480	0.730	0.430

2.3.1 Trust Results (Parts 1 and 2)

Table 2.4 indicates that the participants' trust increased as the week progressed. The data violated the assumption of normality, hence a Wilcoxon Signed Ranks test was used and revealed that there were significant differences between session one and two ($Z = -2.984$, $p = 0.003$) and session one and three ($Z = -2.897, p = 0.004$). There was no significant difference between day two and three.

2.3.2 DBQ Results (Part 1 only)

As shown in Table 2.5, of the participants sampled, errors were the least common driver behaviour (m = 0.61), followed by violations (m = 0.91) and then lapses (m = 1.43). Looking at the standard deviation, violations had the highest variation between the participants (sd = 0.730). The data was not normally distributed; hence the non-parametric Wilcoxon Signed Ranks test was used.

The test reported a significant difference between the lapse and violation scores ($Z = -2.465, p = 0.014$), and the lapse and error scores ($Z = -3.631, p = 0.000$). There was no significant difference observed between the violation and error scores. Lapses were the most common driving mistake the participants made in their driving experience, more so than both violations and errors, which showed no statistical difference between them. According to the aforementioned definitions from [25], it would then be reasonable to expect that lapses were significantly more common than violations or errors.

Table 2.6 DALI results from Part 2

	Measures		
DBQ Results	1	2	3
N	27	27	27
Mean	49.9	45.5	39.5
Std. Dev.	19.6	19.9	19.4

2.3.3 DALI Results (Part 2 only)

The results indicate that the perceived workload decreased as the week progressed (Table 2.6). Data was normally distributed, so a Repeated Measures ANOVA was carried out. The analysis found that between sessions 1 and 2 ($p = 0.028$) and session 1 and 3 ($p = 0.01$) there was a significant drop in the perceived workload. There was no significant difference between session 2 and 3.

2.3.4 Fixations

2.3.4.1 Fixations and Trust (Parts 1 and 2)

Table 2.7 shows the correlation between trust and fixations to the information on the surrogate display.

No significant correlations were found between the fixations to the information icons and trust.

2.3.4.2 Fixations and DBQ (Part 1 only)

Table 2.8 shows the correlation between the driver behaviour questionnaire results and fixations to the information on the surrogate display.

There were significant correlations found between the driver behaviour scores and particular pieces of information. Lapses were found to correlate negatively with fixations towards the navigation information (Figure **??**, Table 2.2) presented on the display ($r = -0.541, p = 0.025$) but positively with road signs ($r = 0.484, p = 0.049$). Violations were found to positively correlate with traffic information ($r = 0.482, p = 0.050$). Errors were found to positively correlate with both traffic information ($r = 0.555, p = 0.021$) and the automated driving indicator ($r = 0.564, p = 0.018$).

2.3.4.3 Fixations and DALI (Part 2 only)

Table 2.9 shows the correlation between the DALI workload measure and fixations to the information on the surrogate display.

Table 2.7 Correlation between trust and fixations

Correlation	Icon	Trust		
		Measure 1	Measure 2	Measure 3
N	–	44	44	44
Action Explanation		0.068	−0.07	0.072
Auto Indicator		−0.103	−0.244	−0.161
Battery		0.073	−0.281	−0.235
Energy Usage		−0.122	−0.06	−0.253
Hazard Scanner		−0.04	0	−0.252
Navigation		−0.049	−0.08	−0.199
Road Signs		−0.144	−0.264	−0.267
Traffic		0.049	−0.078	−0.21
Vehicle Warnings		−0.029	−0.184	−0.075

Table 2.8 Correlation between DBQ and fixations

Correlation	Icon	DBQ		
		Lapses	Violations	Errors
N	–	17	17	17
Action Explanation		0.037	0.035	0.230
Automated Driving Indicator		0.374	0.285	0.564*
Battery Charge Level		0.114	0.224	0.406
Energy Usage	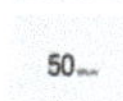	−0.037	−0.228	0.306
Hazard Scanner		−0.152	−0.142	−0.088
Navigation		−0.541*	0.095	−0.296
Road Signs		0.484*	0.298	0.447
Traffic		0.149	0.482*	0.555*
Vehicle Warning		−0.037	−0.228	0.306

Table 2.9 Correlation between DALI and fixations

Correlation	Icon	DALI Measure 1	DALI Measure 2	DALI Measure 3
N	–	27	27	27
Action Explanation		−0.042	0.1	0.096
Automated Driving Indicator		0.27	0.218	0.064
Battery Charge Level		0.221	0.106	0.26
Energy Usage		0.083	−0.121	0.021
Hazard Scanner		−0.024	−0.105	0.042
Navigation		0.136	0.215	−0.013
Road Signs		0.468*	0.236	0.269
Traffic		−0.17	−0.121	0.016
Vehicle Warning		−0.069	−0.108	0.12

There was one significant correlation between the DALI score for the first session and the fixations to the Road Signs information ($r = 0.468, p = 0.014$). All other correlations were non-significant.

2.3.5 Between trust, DBQ and DALI

2.3.5.1 Trust and DBQ (Part 1 only)

The goal was to understand if a driver's driving behaviours in vehicles today would affect their predisposition to trust an automated system. This correlation was only applicable to Part 1 of the study. The results are presented in Table 2.10.

The results found three significant interactions. On the first day, trust had a significant negative correlation with the lapse ($r = -.551, p = 0.022$) and error scores ($r = -0.526, p = 0.030$). On the fifth day, trust correlated negatively with the lapse score ($r = -0.487, p = 0.047$). The results would suggest that on the first day of using a partially automated system, the participants who had experienced more lapses and errors in their driving, were less likely to trust the automated system.

Table 2.10 Correlation between driver behaviour and trust

	DBQ		
Correlation	Lapses	Violations	Errors
N	17	17	17
Trust 1	−0.551*	−0.230	−0.526*
Trust 2	−0.360	−0.039	−0.449
Trust 3	−0.487*	−0.165	−0.410

Table 2.11 Correlation between DALI and trust

Correlation	DALI
N	27
Trust 1	−0.434*
Trust 2	−0.413*
Trust 3	−0.450*

2.3.5.2 Trust and DALI (Part 2 only)

The goal of this correlation was to understand the relationship between the trust a participant places in the partially automated vehicle and their perceived workload. This correlation was only applicable to Part 2 of the study. The results are presented in Table 2.11.

The results found that trust had a significant negative correlation with the DALI perceived workload measure on all days (Trust 1, $r = -0.434, p = 0.024$; Trust 2, $r = -0.413, p = 0.032$; Trust 3, $r = -0.450, p = 0.018$).

2.4 Discussion

This two part study aimed to characterize the participants' based on their trust in partially automated vehicles, driving behaviour and perceived workload to understand if this affected the information they used. Information was categorized according to the SRK model to ensure the information presented during the study was cognitively balanced and representative.

2.4.1 Fixations

2.4.1.1 Fixations and Trust (Parts 1 and 2)

The results indicated that there was no relationship between the level of trust in the partially automated system and fixations to any particular piece of information on the interface. The implication here is that no single piece of information can be relied on to influence driver trust in partially automated vehicles. However, trust did significantly increase after the first day, but it would appear that this was a result of the steady–state performance of the vehicle and not the information presented to them. The key takeaway of this result is that whether a participant exhibits a high or low trust does not appear to affect their usage of information inside a partially automated vehicle during steady–state driving. Rather, it would appear that increasing exposure is the most effective method of trust calibration.It may come down to individual preferences as to how information is utilized and acted upon[14], but it would be inaccurate to base any information adaptation or design decisions on the trust level of a participant for steady–state driving.

An important next step in this area would be to understand the effect of an emergency scenario on trust and if there are particular pieces of information that are then relied on by drivers to recalibrate any consequent change in trust.

However, it is important to note that trust is still an important measure in the design of partially automated vehicles as a whole, as discussed later in this paper.

2.4.1.2 Fixations and DBQ (Part 1 only)

The results only found positive correlations between skill based information elements of the display and each of the three aspects of the DBQ. The other significant negative correlation was the Navigation $(r = -0.541, p = 0.025)$ (a rule based information); meaning a participant with a higher lapse score, used Navigation more. This could be because they felt the skill–based information was able to meet their information requirements and Navigation became redundant. Skill–based information like the Automated Driving Indicator can provide a simple confirmatory icon to tell the driver everything is working as it should, without the driver needing to compare the Navigation information with what the vehicle is actually doing.

Looking more generally at the DBQ, those drivers who scored higher appear to rely more on skill–based information,i.e information that is more intuitive and requires less cognitive processing. If there was a way to categorize drivers and identify those who are more prone to error as measured by the DBQ, this could be a good case for adapting the information provided to those drivers with more skill based, less cognitively demanding information.

2.4.2 Fixations and DALI (Part 2 only)

With respect to the DALI and fixations, only the Road Signs on the first day of trials showed a significant effect $(r = 0.468, p = 0.014)$. This meant that the participants who used the Road Signs information more, subjectively reported a higher perceived workload.Raw fixation results would indicate the Road Sign information was commonly used by the participants to monitor the road condition and confirm the vehicle's sensors were accurately perceiving the road ahead. This would then suggest that the monitoring task does cause perceived workload to increase, confirming this previously observed effect [30, 31].

Furthermore, although information was categorized according to cognitive demand using the SRK model [16], the prototyping and subsequent redesign of icons was intended to balance the visual saliency and hence achieve an equitable cognitive demand across all the information. This may explain why only one piece of information was correlated with perceived workload.

2.4.3 Between Trust, DBQ and DALI

The final aspect of analysis was to compare the Jian trust rating with both the DBQ and the DALI. In the previous section, the results suggested that trust was not a driving factor behind the information used inside a partially automated vehicle. However, it remains an important factor in other aspects as discussed here.

2.4.3.1 Trust and DBQ (Part 1 only)

Between the DBQ and Jian Trust, the results found significant negative correlations between trust and lapses on days one and three (r-day $1 = -0.551$, $p = 0.022$; r-day $2 = -0.487$, $p = 0.047$). The error scores displayed a significant correlation with trust ($r = -0.526, p = 0.030$) but only on day one. These results indicate that drivers who are more prone to errors and lapses, have a lower propensity to trust the automated system.Notably, drivers more prone adverse driving behaviour are also those who are more likely to benefit from the enhanced safety features of partially automated vehicles. However, these results indicate that those drivers are less likely to trust the system and hence not use it.This lack of trust could also be indicative of the driver showing awareness of their own shortcomings when the vehicle enters automated driving mode and consequently not trusting the system. However, it is interesting to note how the correlations between trust and lapses/errors became weaker as the week progressed. This would suggest that the participants' previous driving experiences became less relevant to their use of the automated vehicle with increased exposure.

2.4.3.2 Trust and DALI (Part 2 only)

The final aspect was to compare the DALI workload rating against trust. The results found a significant negative correlation on all days indicating that as trust increased in the partially automated system, the perceived workload went down. This effect has been observed before [10, 11, 32, 33] and the results from this study confirm the effect that if mental workload is reduced, then it is likely trust in the automated system will increase in partially automated vehicles. Of all the factors investigated in this study, workload had the biggest effect on the level of trust drivers placed in the system. However it should be noted that the goal should not be to minimize workload as under loading can reduce situational awareness and have a negative impact on how safely the partial automation is used [34]. On the contrary, too high a workload can reduce the experience and safety of the vehicle [35]. Comparing this to the fixation behaviour where there was no correlation between workload and information use, this suggests there needs to be more investigation into how cognitive workload may be more appropriately managed.

2.5 Conclusion

This study used a unique longitudinal within-subjects design to test how trust, driver behaviour and perceived workload relate to information requirements inside partially automated vehicles. This study found that trust increased significantly by the end of the trial week. Conversely, workload significantly decreased by the end of the week.This study has also shown how trust is more influenced by increasing exposure to a partially automated system than the information presented, as a participant's trust level was found to have no correlation to information fixations. However, that is not to say that trust is not an important measure as it was found to significantly negatively correlate with the perceived workload experienced by drivers. Increased workload may lower the user experience of the automated system, hence, trust remains an important measure. Drivers who may be more prone to adverse driving behaviour may rely more on skill–based information. These drivers are also less likely to trust a partially automated vehicle. Further, the DBQ has shown to be potentially effective at predicting a driver's propensity to trust a partially automated vehicle, but the correlation strength of this relationship decreases with increasing exposure to the automated system.

This work has raised important implications for the design of future interfaces in automated vehicles. Namely, the effect of behavioural factors such as trust, workload and information usage have been shown to play a role in how the system is used. As supported by this research, future work must consider looking at how information presented inside partially automated vehicles can be adapted to match these driver trust and behaviour profiles, and also consider how these information requirements change over time.

References

[1] P. J. Muller, "Driverless Transportation—Two Future Scenarios," in *International Conference on Transportation and Development 2016*, 2016, pp. 140–151.

[2] T. B. Sheridan, "Human centered automation: oxymoron or common sense?," in *Proceedings of the IEEE International Conference on Systems, Man and Cybernetics*, 1995, vol. 1, pp. 823–828.

[3] D. B. Kaber and M. R. Endsley, "The effects of level of automation and adaptive automation on human performance, situation awareness and workload in a dynamic control task," *Theor. Issues Ergon. Sci.*, vol. 5, no. 2, pp. 113–153, Jan. 2004, doi: 10.1080/1463922021000054335.

[4] M. T. Dzindolet, S. A. Peterson, R. A. Pomranky, L. G. Pierce, and H. P. Beck, "The role of trust in automation reliance," *Int. J. Hum. Comput. Stud.*, vol. 58, no. 6, pp. 697–718, Jan. 2003, doi: 10.1016/S1071-5819(03)00038-7.

[5] V. A. Banks, A. Eriksson, J. O'Donoghue, and N. A. Stanton, "Is partially automated driving a bad idea? Observations from an on-road study," *Appl. Ergon.*, vol. 68, pp. 138–145, Apr. 2018, doi: 10.1016/j.apergo.2017.11.010.

[6] J. K. Choi and Y. G. Ji, "Investigating the Importance of Trust on Adopting an Autonomous Vehicle," *Int. J. Hum. Comput. Interact.*, vol. 31, no. 10, pp. 692–702, Jul. 2015, doi: 10.1080/10447318.2015.1070549.

[7] K. E. Schaefer, J. Y. C. Chen, J. L. Szalma, and P. A. Hancock, "A Meta-Analysis of Factors Influencing the Development of Trust in Automation: Implications for Understanding Autonomy in Future Systems," *Hum. Factors*, vol. 58, no. 3, pp. 377–400, Mar. 2016, doi: 10.1177/0018720816634228.

[8] A. Ulahannan et al., "Using the ideas café to explore trust in autonomous vehicles," in *Advances in Intelligent Systems and Computing*, 2019, vol. 796, pp. 3–14, doi: 10.1007/978-3-319-93888-2_1.

[9] V. A. Banks, K. L. Plant, and N. A. Stanton, "Driver error or designer error: Using the Perceptual Cycle Model to explore the circumstances

surrounding the fatal Tesla crash on 7th May 2016," *Saf. Sci.*, vol. 108, pp. 278–285, Oct. 2018, doi: 10.1016/j.ssci.2017.12.023.

[10] N. Lyu, Z. Duan, L. Xie, and C. Wu, "Driving experience on the effectiveness of advanced driving assistant systems," in *2017 4th International Conference on Transportation Information and Safety, ICTIS 2017 – Proceedings*, 2017, pp. 987–992, doi: 10.1109/ICTIS.2017.8047889.

[11] U. E. Manawadu, T. Kawano, S. Murata, M. Kamezaki, and S. Sugano, "Estimating driver workload with systematically varying traffic complexity using machine learning: Experimental design," in *Advances in Intelligent Systems and Computing*, 2018, vol. 722, pp. 106–111, doi: 10.1007/978-3-319-73888-8_18.

[12] W. Piechulla, C. Mayser, and H. Gehrke, "Reducing driver's mental workload by means of an adaptive man-machine interface," vol. 6, no. 4, pp. 233–248, Dec. 2003.

[13] S. Khastgir, S. Birrell, G. Dhadyalla, and P. Jennings, "Calibrating trust through knowledge: Introducing the concept of informed safety for automation in vehicles," *Transp. Res. Part C Emerg. Technol.*, vol. 96, pp. 290–303, Jan. 2018, doi: 10.1016/j.trc.2018.07.001.

[14] A. Ulahannan et al., "User expectations of partial driving automation capabilities and their effect on information design preferences in the vehicle," *Appl. Ergon.*, vol. 82, p. 102969, 2020, doi: https://doi.org/10.1016/j.apergo.2019.102969.

[15] A. Ulahannan, P. Jennings, L. Oliveira, and S. Birrell, "Designing an Adaptive Interface: Using Eye Tracking to Classify How Information Usage Changes Over Time in Partially Automated Vehicles," *IEEE Access*, vol. 8, pp. 16865–16875, 2020, doi: 10.1109/ACCESS.2020.2966928.

[16] J. Rasmussen, "Skills, Rules, and Knowledge; Signals, Signs, and Symbols, and Other Distinctions in Human Performance Models," *IEEE Trans. Syst. Man Cybern.*, vol. SMC-13, no. 3, pp. 257–266, Jan. 1983, doi: 10.1109/TSMC.1983.6313160.

[17] J. Michon, "A critical view of driver behavior models: what do we know, what should we do?," *Aust. J. Prim. Health*, vol. 21, no. 1, pp. 485–524, Jan. 1985.

[18] G. Geiser, "Man Machine Interaction in Vehicles," *Atz*, vol. 87, no. 74–77, pp. 74–77, Jan. 1985.

[19] S. Birrell, M. Young, N. Stanton, and P. Jennings, "Using adaptive interfaces to encourage smart driving and their effect on driver workload," in

Advances in Intelligent Systems and Computing, vol. 484, no. Chapter 3, Cham: Springer International Publishing, 2017, pp. 31–43.

[20] D. B. Kaber, J. M. Riley, K.-W. Tan, and M. R. Endsley, "On the design of adaptive automation for complex systems," *Int. J. Cogn. Ergon.*, vol. 5, no. 1, pp. 37–57, Jan. 2001.

[21] BSI, "Road vehicles. Ergonomic aspects of transport information and control systems. Specifications and test procedures for in-vehicle visual presentation," BS EN ISO 15008:2017, 2017.

[22] UNECE, "UN Regulation No. 121 – Rev.2 – Identification of controls, tell-tales and indicators," 121 Amend. 7, 2018.

[23] S. I. Becker, A. M. Harris, D. Venini, and J. D. Retell, "Visual search for color and shape: When is the gaze guided by feature relationships, when by feature values?," *J. Exp. Psychol. Hum. Percept. Perform.*, vol. 40, no. 1, pp. 264–291, Jan. 2014.

[24] J.-Y. Jian, A. M. Bisantz, and C. G. Drury, "Foundations for an empirically determined scale of trust in automated systems," *Int. J. Cogn. Ergon.*, vol. 4, no. 1, pp. 53–71, Jan. 2000.

[25] D. Parker, J. T. Reason, A. S. R. Manstead, and S. G. Stradling, "Driving errors, driving violations and accident involvement," *Ergonomics*, vol. 38, no. 5, pp. 1036–1048, Oct. 2007.

[26] A. Pauzié, "A method to assess the driver mental workload: The driving activity load index (DALI)," *IET Intell. Transp. Syst.*, vol. 2, no. 4, pp. 315–322, Jan. 2008.

[27] E. C. Poulton, "Peripheral vision, refractoriness and eye movements in fast oral reading.," *Br. J. Psychol.*, vol. 53, no. 4, pp. 409–419, Nov. 1962.

[28] T. A. Salthouse and C. L. Ellis, "Determinants of eye-fixation duration," *Am. J. Psychol.*, pp. 207–234, Jan. 1980.

[29] J. L. Orquin and K. Holmqvist, "Threats to the validity of eye-movement research in psychology," *Behav. Res. Methods*, vol. 50, no. 4, pp. 1–12, Jul. 2018, doi: 10.3758/s13428-017-0998-z.

[30] J. Stapel, F. A. Mullakkal-Babu, and R. Happee, "Automated driving reduces perceived workload, but monitoring causes higher cognitive load than manual driving," *Transp. Res. Part F Traffic Psychol. Behav.*, vol. 60, pp. 590–605, Jan. 2019, doi: 10.1016/j.trf.2018.11.006.

[31] G. H. Walker et al., "From ethnography to the east method: A tractable approach for representing distributed cognition in air traffic control," *Ergonomics*, vol. 53, no. 2, pp. 184–197, Jan. 2010, doi: 10.1080/00140130903171672.

[32] C. J. Pearson, A. K. Welk, and C. B. Mayhorn, "In Automation We Trust? Identifying Varying Levels of Trust in Human and Automated Information Sources," *Proc. Hum. Factors Ergon. Soc. Annu. Meet.*, vol. 60, no. 1, pp. 201–205, Jan. 2016.

[33] I. J. Reagan and J. P. Bliss, "Perceived mental workload, trust, and acceptance resulting from exposure to advisory and incentive based intelligent speed adaptation systems," *Transp. Res. Part F Traffic Psychol. Behav.*, vol. 21, pp. 14–29, 2013, doi: 10.1016/j.trf.2013.07.005.

[34] J. J. Blum and A. Eskandarian, "Managing effectiveness and acceptability in intelligent speed adaptation systems," *IEEE Conf. Intell. Transp. Syst. Proceedings*, ITSC, pp. 319–324, Jan. 2006, doi: 10.1109/itsc.2006.1706761.

[35] J. M. Hoc, M. S. Young, and J. M. Blosseville, "Cooperation between drivers and automation: Implications for safety," *Theor. Issues Ergon. Sci.*, vol. 10, no. 2, pp. 135–160, Jan. 2009, doi: 10.1080/14639220802368856.

3

A CNN Approach for Bidirectional Brainwave Controller for Intelligent Vehicles

Armando Astudillo Olalla and Fernando García Fernández

Intelligent Systems Lab – UC3M
E-mail: aastudil@ing.uc3m.es; fegarcia@ing.uc3m.es

Our brain is the main factor for the accomplishment of any task in the daily life of the human being, therefore, the patterns of mental thought, which are represented in the neuronal signals, can be used to improve the life of both healthy and disabled users.

Brain–Computer Interfaces (BCI) are used for the reading of brain waves. These sensors allow communication between a computer and the human brain. With them, brain activity readings are obtained to analyse patterns of electrical activity, or elecroencephalogram (EEG), that reflect certain brain orders.

In this work we review some of the most sophisticated and relevant techniques in the classification of these patterns to control. We propose a new method of classifying brain patterns with the use of a low-cost helmet and convolutional neural networks (CNN). The results obtained are very significant since on the one hand they demonstrate the improvement of the prediction with this type of networks, and on the other they indicate us that it is preferable to train a model by subject, although not completely necessary.

3.1 Introduction

In the last few years, the technology applied to improve vehicle safety and driving assistance has advanced enormously, reaching the stage of new models capable of driving certain routes without a driver managing the controls.

These developments are carried out in parallel with the implementation of new technologies that allow people with different disabilities to drive, such as special controls for people who cannot use the pedals of the vehicle.

This work aims to provide an efficient (both in computational cost and monetary cost) and effective solution to implement a brainwave controller for route decision in a vehicle with automated control.

Brain–Computer Interfaces (BCI) has the potential to enable severely disabled people to drive different machines such as computers or manipulators by brain activity rather than physical means. Many technological advances have occurred recently towards developing devices which allow people to use some machine through muscle control, which may not be useful for totally paralysed people. Therefore, BCI methods are the main alternative for this people to communicate with any electronic device.

The BCI is based on the detection of brain electrical activity, produced by brain waves, in order to control a machine or device with thoughts after a certain calibration. Electroencephalography (EEG) measures voltage fluctuations resulting from ionic current flows within the neurons of the brain. The first International Meeting on BCI technology took place in Renserlaerville (NY), in June 1999 and was organised by the BCI research team at the Wadsworth Center of the NY State Department of Health and State University of New York. Different approaches to the training of subjects in the control of EEG signals, and some techniques to record these waves were discussed.

3.1.1 Human Brain

The study of the human brain has been one of the fields of research that have most attracted the attention of scientists for thousands of years. It is known that the Egyptians, more than 5,000 years ago, made the first discoveries in the field of neuroscience, practised diagnoses related to neurological conditions and performed simple operations. Since then, medical techniques have improved considerably, and today, other sciences such as engineering and physics are interested in the functioning of the brain and its possible practical applications.

The human brain is divided into different lobes, each lobe is specialised in certain activities such as the use of memory, motor activity or the resolution of logical problems.

- **Frontal Lobe**: It is in charge of cognitive task, maintaining the necessary attention for solving problems and planning tasks.

- **Parietal Lobe**: It deals with the visual control and tactile perception by controlling some variables such as temperature and pressure.
- **Temporal Lobe**: Various tasks are associated with it, including the ability of long-term memory, emotional responses and auditory perception.
- **Occipital Lobe**: It is in charge of processing visual stimuli, giving shape and colour to objects and detecting movement.

3.1.2 Brainwaves Features

The brain is the organ in charge of processing and executing the signals of the central nervous system. This system collects information from different nerves and is transmitted through neurons to the brain, which analyses it and sends various orders through other neurons. Communication between the neurons that make up this system is carried out through synapses, a biological process by which electrical activity (EEG, Figure 3.1) is generated from the reaction of chemical substances produced by the emitting neuron.

The first advance made in the construction of the brain map came from the hand of the German neurologist Hans Berger, who in 1924 deduced the existence of brain waves, soon after managed to prove his theory by electrodes inserted in the cerebral cortex of a subject with which it was possible to detect electrical oscillations in the brain, thus creating the first human encephalogram (EEG, Figure 3.2).

Since this first discovery on the cerebral waves, the relation between diverse actions like the resolution of mathematical operations or to sleep, and the cerebral activity was studied.

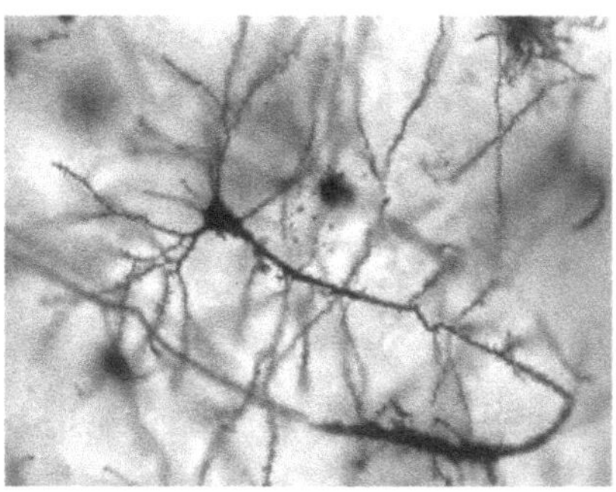

Figure 3.1 Golgi stained pyramidal neuron in the hippocampus of an epileptic patient. 40 times magnification. (Source: MethoxyRoxy CC).

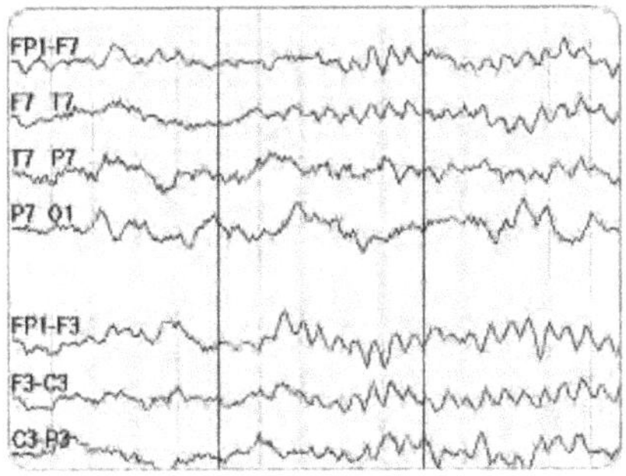

Figure 3.2 Electrical signals generated by brain activity. (Source: About Kids Health).

Brainwaves are classified depending on which frequency they are modulated:

- **Delta**: 0.5–2.75 Hz
- **Theta**: 3.5–6.75 Hz
- **Low Alpha**: 7.5–9.25 Hz
- **High Alpha**: 10–11.75 Hz
- **Low Beta**: 13–16.75 Hz
- **High Beta**: 18–29.75 Hz
- **Low Gamma**: 31–39.75 Hz
- **Mid Gamma**: 41–49.75 H

These waves relate to movement and the level of attention in a certain task [1].

3.1.3 BCI Research

In 2001, Guger et al. developed a BCI model to classify EEG patterns while subjects thought of the left- and right-hand movements[2], this model has an accuracy between 70% and 95%. An adaptive auto-regressive model (AAR) [3] and linear discriminant analysis (LDA) [4] were developed. It was tested with a BCI helmet covering the motor and somatosensory zones (C3 and C4 of 10-20 Standard Disposition, shown in Figure 3.3). In 2003, they also published a data set for a contest to develop a model capable of recognising brain patterns, to be tested by BCI headsets[5].

In 2008, Xu et al. combined the discrete wavelet transformation (DWT) with auto-regressive model to analyse the Graz data set, they achieved the classification accuracy of 90% [6]. The method has 6 statistical wavelet coefficients and 6 AAR coefficients for each channel, giving a total of 24

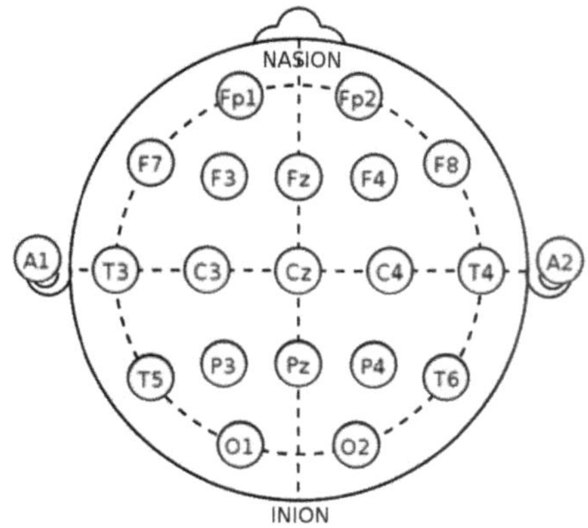

Figure 3.3 10–20 Standard Disposition. (Source: Christopher B. under CC License).

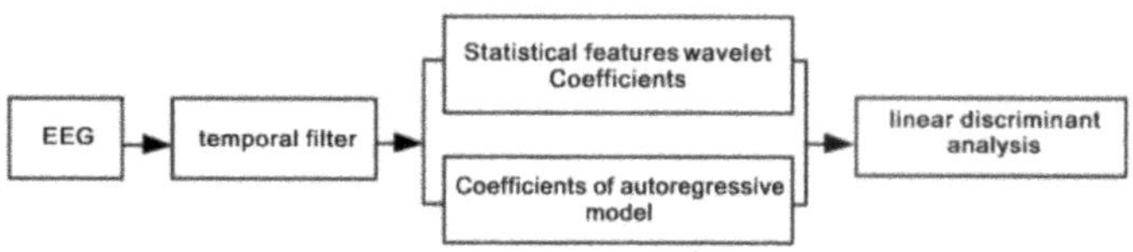

Figure 3.4 Flow chart of the data processing. (Source: Xu in [6]).

features for a motor imagery task. These parameters were selected as inputs of LDA classifier. The results show the Daubechies order 10 gave the best performance and the recognition rate is as high as 90.0%. The results indicate that method of combining DWT with AR model is capable of extracting more useful information from the simultaneously acquired motor imagery EEG.

One of the most promising experiments in this field was developed by the DARPA (Defense Advanced Research Projects Agency) at the end of 2012 [7]. A group of researchers managed to get a middle-aged quadriplegic woman to move a robotic arm with great precision. In order to do this, they implanted two 96-channel intracortical sensors (Figure 3.5, up). The information that this device sends to the computer is immense and after approximately 13 weeks, the subject was able to handle the arm in any spatial direction, and even grab and handle simple objects.

This work proposes a new approach based on the work developed in [8], which aims to provide a solution to decide the direction in an automated vehicle using neural networks and a cost-efficient BCI helmet.

The methodology followed in this study suggests the use of neural networks for the estimation of the direction to which it is intended to go.

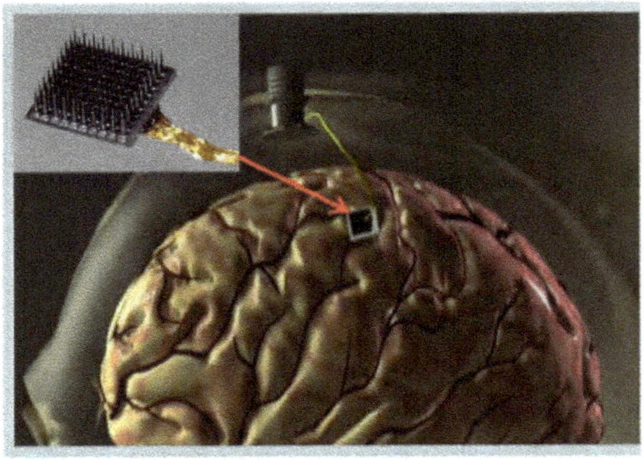

Figure 3.5 Up: 96-channel intracortical sensor. Down: Functional magnetic resonance imaging. (Source: in [7]).

The configuration of the network consists of 7 neurons for the first layer, 3 for the second and 2 neurons for the third and last layer. The experiments were carried out on a wide group of subjects, of different ages and gender.

For the training tests the data set is divided into 80% to train the network, and 20% for validation. The results obtained for this first experiment show some hope for the possibility of its use as a sense estimator, since 83.33% of success was obtained when people thought of going to the left, and 75.00% of success when it was required to go to the right.

For the validation tests a total of 50 samples were collected, the algorithm was able to recognise 63% of the cases correctly, which means that it was not a fully robust system.

These results may be due to changing conditions between simulation and field tests, as well as the subjects' own perception of right and left.

That is why two new approaches have been developed: the first uses the same methodology as in [8] but treating each individual independently, without mixing the wave patterns of the group, on the other hand a new methodology is made based on the CNN to solve both the problem of classification in group and individual. In summary, our key contributions are: (a) a comparison between the general method developed in [8] and an individual method using the same architecture; (b) a novel CNN-based approach for both individual and general data set and (c) a fair comparison between these two architectures.

The work presented in this paper is divided into several sections: Section 2 explains the hardware system used, Section 3 describes the processing of

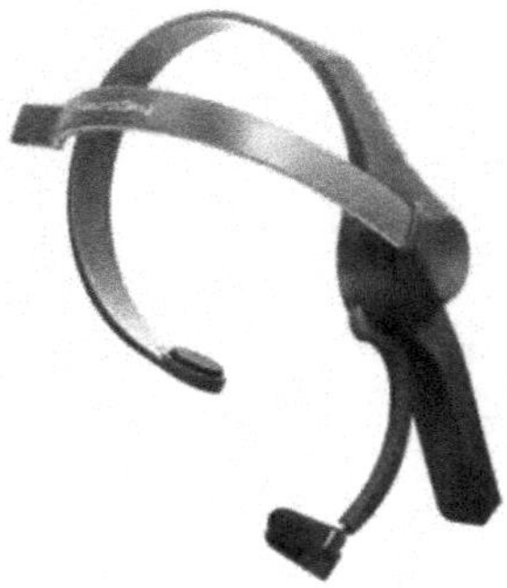

Figure 3.6 Mindwave Neurosky Sensor [9].

the data and the methodology used for classification, Section 4 presents the experiments carried out, as well as the results obtained and to conclude, a brief conclusion and proposals for future work are presented in Section 5.

3.2 Setup Overview

This section describes the hardware system used to obtain the brainwave data and the vehicle in which the system has been tested.

3.2.1 Brainwave Sensor

A low-cost sensor developed by Neurosky is used to read brain waves. The Mindwave Mobile (Figure 3.6) is a brainwave headset that can send readings via Bluetooth and contains two sensors: one in the left ear for signal reference and another located on the forehead for reading the waves. To obtain the amplitude of the different brain waves, the device has a microchip that processes the raw signal by using a Fast Fourier Transform (FFT). The brainwaves amplitudes oscillate from 0 to 255 [9].

This sensor has a sampling rate of 1.0 [Hz], it means that the system is going to need some seconds to have an output. The need of a better sensor to obtain a faster response is discussed in section 5.

3.2.2 Vehicle Platform

To carry out the experiments on a real platform, an electric golf cart is used, which is equipped with multiple sensors on board, such as the LIDAR sensor, stereo camera, GPS and control over its movement. Figure 2 shows

Figure 3.7 Research platform "iCab".

this vehicle, which is part of the Intelligent Campus Automobile ('iCab') project [10] [11].

3.3 Methodology

3.3.1 Data Reading

First step is to acquire the data from the Mindwave Mobile sensor. In order to do this, we use the ROS driver developed in [12], modifying the output to obtain the type of waves needed, since the initial driver just provide us the attention and meditation level without the values of the waves.

A Python script has been coded; hence, the ThinkGear Communications Protocol must be used to decode signals' values. This communication protocol is used to connect the sensor to the PC via Bluetooth. Thus a message is obtained with the 8 basic brain waves, in packages of 24-bytes each.

3.3.2 Data Filtering

Once the waves have been decoded, these data must be filtered. This is an essential step to eliminate repeated samples (Figure 3.8), as it is sampled at a frequency greater than twice the sending frequency.

3.3.3 Input Processing

Once filtered (Figure 3.9), the data has to be converted to the input format of the classifier. Part of the acquired waves, as well as variations of these waves,

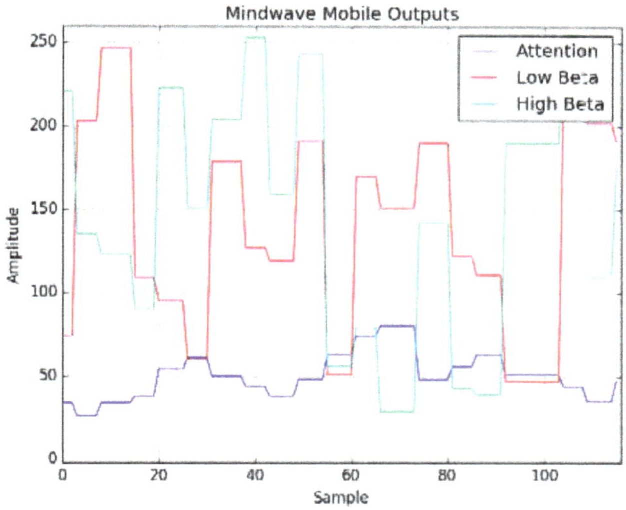

Figure 3.8 Example of raw data obtained from the brainwave sensor.

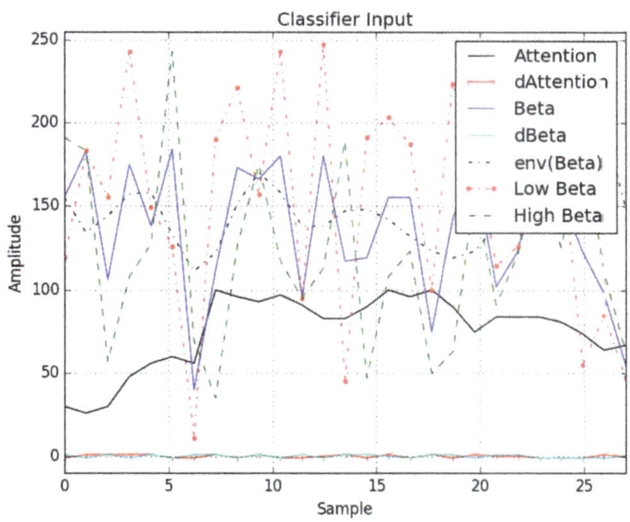

Figure 3.9 Neural Net input example.

are converted into the specific format of each classifier, which is specified in the methodology of these classifiers. Both approaches use the following as inputs: attention level, attentions derivative, beta wave, beta wave derivative, beta wave envelope and low and high beta waves.

To obtain the derivatives of the waves ($d(W_x)$), the behaviour of the entire wave is analysed to determine when it is increasing or decreasing, according to:

$$d(W_x) = \begin{cases} 1, & \text{if } W_x(t) < W_x(t+1) \\ 0, & \text{if } W_x(t) = W_x(t+1) \\ -1, & \text{if } W_x(t) > W_x(t+1) \end{cases} \tag{3.1}$$

The values obtained by the sensor for low beta and high beta waves are interpolated with a cubic interpolation algorithm as it provides a correct adjustment with low computational consumption. On the other hand, the envelope curve of the waves is calculated and the average is calculated to obtain a single beta wave:

$$Env_{\mathbf{Beta}}(t) = \frac{EnvLowBeta(t) + EnvHighBeta(t)}{2} \tag{3.2}$$

3.3.4 NN Classifier

The implementation is similar to the one developed in [8], this method consists in the use of neural networks that are modelled as a multi-layer perceptron based on Quasi-Newton optimisation for the logarithmic cost function described in:

$$V\left(f(x), y\right) = -y \cdot ln(f(x) - (1-y) \cdot ln(1 - f(x)) \tag{3.3}$$

The activation function for the hidden layers of the implemented networks is the Rectified Linear Unit Function (RELU):

$$f(x) = max(0, x) \tag{3.4}$$

The structure of the neural network consists of an input layer, three hidden layers and an output layer with a single neuron. This network is modelled using Scikit-Learn, a Machine Learning library for Python [13]. The proposed algorithm consists of four different neural networks, two for each direction. The output of each classifier generates two classes:

- C_0: if the subject is not sufficiently concentrated.
- C_1 : if the level of attention is sufficient to be thinking about the motion.

Furthermore, each classifier returns one output for each instance of reading, however, to obtain a robust result, a reading of several samples is required. Thus, the output of each classifier is predicted by comparing, for each sample obtained, the probability of belonging to a certain class (3.5) and the error ratio (3.6), which depends directly on the level of attention:

$$Distance_{sense} = \sum_{i=0} P(x_i|C_0) - P(x_i|C_1) \tag{3.5}$$

$$Error_{sense} = \frac{Number_{failures}}{Number_{samples}} \tag{3.6}$$

The final result has three possible states:

1. NULL: the result cannot be classified.
2. LEFT: the subject is thinking left.
3. RIGHT: the subject is thinking right.

3.3.5 CNN Classifier

Instead of solving the classification problem by treating the data as discrete as in [8], a CNN architecture developed in PyTorch will be used. The values of each sampled variable reach values ranging from 0 to 255, so they can be processed as if they were pixels of a greyscale image, except for the derivative values that can be interpreted as binary pixel values. Taking into account this new form of data, it can be fed as an input for a CNN network, so this input has the form of an image of $1 \times 20 \times 7$, 20 for the number of samples per session and 7 for the number of waves sampled.

In this paper, we are not going to use a deep CNN, as the purpose is to create a low-cost computational solution for a low-cost sensor. Therefore, two different architectures have been tested: MindNet_1 and MindNet_2.

3.3.5.1 MindNet_1

The first CNN-based solution consists of a simple network (Figure 3.10) composed of a $1 \times 20 \times 7$ input, a 2D convolutional layer, a max pool layer, followed by two fully connected layers: the first of 540×64 and the second of 64×2, and a Log-Softmax layer (3.7) for classification.

$$LogSoftmax(x_i) = log\left(\frac{exp(x_i)}{\sum_j exp(x_j)}\right) \tag{3.7}$$

A NLLLoss (3.8) cost function has been used for parameter optimisation, whereby the loss tends to -1. It is a function for the classification of problems with 'C' classes, the entry for the function is a tensor containing

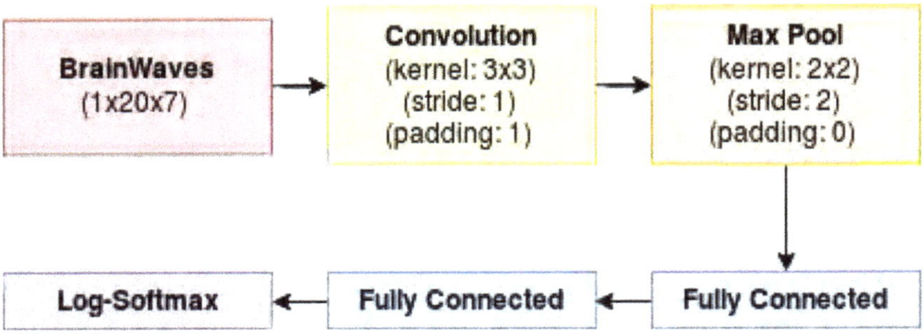

Figure 3.10 MindNet_1 CNN-Architecure.

the logarithmic values of the probability of belonging to each class:

$$l(x, y) = \sum_{n=1}^{N} \left(\frac{l_n}{\sum_{n=1}^{N} w_{y_n}} \right) \tag{3.8}$$

An AdaDelta model [14] has been used as an optimiser with an epoch equal to 200.

3.3.5.2 MindNet_2

Due to new studies carried out on the use of batch normalization layers in which it is stated that each neuron can learn certain characteristics more independently of the result of the others [15], and that brain patterns can be difficult to identify, it has been decided to implement these layers in a new neuronal architecture (Figure 3.11). This network consists of an input of 1x20x7, followed by two blocks composed of a convolutional layer and a BatchNorm layer, then go through a module fully connected without Log-Softmax layer because as a cost function is used the Cross Entropy Loss function that combines a Log-Softmax layer (3.7) next to the NLLoss equation (3.8).

AdaDelta model is used again as optimiser with an epoch of 200.

3.4 Experimental Works and Results

This section presents the experiments carried out both for the general classifier and for the individual classifiers, as well as the results obtained.

These experiments allow us to create some data sets, presented in Table 3.1, so that various approaches can be tested on them.

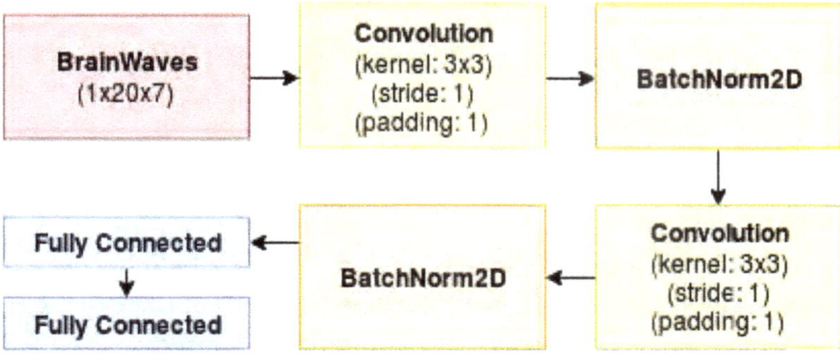

Figure 3.11 MindNet_2 CNN-Architecure.

Table 3.1 Generated data set

Data set	Samples	Subjects	Age Range
General	481	34	21–50
Individual	576	2	19–22

Figure 3.12 Animated arrow scene.

3.4.1 General Classifier

As described in the work [8], the experiment carried out consists of obtaining a single classifier for a varied group of people.

To obtain the data set, several subjects are required to wear the BCI helmet to read their brain waves while thinking in the direction (left or right) followed by an animated arrow, displayed on a screen, while pressing a key with their hand in the direction of the arrow. This process takes a minute and a half and takes place in both directions. For this experiment, the subject must be focused on moving the arrow with his mind, without knowing that he is moving by himself. Thus, the thought is reinforced in only one of the senses.

To make it easier for the subjects to concentrate on their thoughts about moving to the left or right, they are shown an arrow in an HTML application (Figure 3.12) that moves in one direction or the other while pressing a key with the hand corresponding to the direction of the arrow.

Table 3.2 Results for general classifier

Classifier	Train Score	Real Test Score
NN	79,17%	63,5%
MindNet_1	65,00%	70,00%
MindNet_2	**75.00%**	**70,00%**

A total of 481 samples were obtained from 34 subjects between the ages of 21 and 50, of different genders, both left-handed and right-handed. Once the data is obtained, it is divided into 80% for network training and 20% for validation.

To test the success of the classifiers, more experiments were conducted using the 'iCab' platform. They were carried out in an uncontrolled environment where subjects thought of the route the vehicle should take. Totally, 51 new samples to be classified were obtained, distributed in such a way that they can be used in the three classifiers presented.

The results obtained are shown in Table 3.2. This table indicates how the precision of prediction increases when using convolutional networks as opposed to the use of simple neural networks. However, it is not a good enough approach to be implemented as a real solution, so an individual data set will be tested.

3.4.2 Individual Classifier

In spite of having improved the classification of the sense in which a subject thinks, a new experiment is carried out to test whether training a model for each subject reaches a level of prediction that can be really useful.

For consistent training data, the experiments were carried out on different days and under different conditions: in an empty room and a room with people talking around. No other applications such as the animated arrow from the previous experiment (Figure 3.12) were needed to induce concentration on the subject, since, being a single person, he or she must focus on one sense or another depending on his or her sensations. In addition to these new conditions, the session time has been reduced to 30 s, instead of the minute and a half of the previous experiment, this is done in consideration of decreasing the time that the subject has to be concentrated, since the longer the exposure time, the greater the fatigue and the greater the distraction.

The results of this experiment have produced a total of 576 samples, which, as in the previous experiment, are divided into 80% for training and 20% for validation.

Table 3.3 Results for individual classifier

Classifier	Train Score	Real Test Score
NN	90,00%	83,33%
MindNet_1	87,00%	93,00%
MindNet_2	**85,17%**	**98,00%**

Table 3.4 Computational time

Data	Acquisition Time [s]	NN [ms]	CNN [ms]
General	90,00	727,44	**2,62**
Individual	**30,00**	720,68	**2,57**

The success levels for an individual classifier show that training a model independently for each person is an improvement to be taken into account.

The results obtained show how the system is capable of correctly classifying a high percentage of thoughts, equivalent to the results of the work described earlier. These results demonstrate that the use of convolutional networks instead of basic networks allow to extract better characteristics to detect the pattern of each individual.

3.4.3 Computational Time

Two metrics have been compared to obtain the real computational time cost: the time needed to acquire the data and the inference time, it means the time needed to process the data obtained and give a result.

It is shown that the use of the novel CNN approach with a model for each subject is a faster solution, reducing significantly the calculation time in comparison to the basic NN approach [8].

3.5 Conclusion and Future Work

This paper reviews some of the techniques used in the BCI field. All of them use very expensive sensors with multiple channels that map brain activity with good resolution.

This study develops a new methodology for predicting route decisions with low-cost BCI sensors. Three different approaches are compared: the first based on the work done in [8] and the other two based on the use of convolutional networks.

The use of a convolutional architecture to classify the samples collected by a single sensor as if they were images gives better results (**98.00%**

accuracy, Table 3.3) in a shorter time (Table 3.4), thus increasing the robustness of the systems to real levels of usability and close to much more expensive systems.

For future work it is proposed to use a different helmet, which has more sensors and a faster sampling rate, in order to be integrated optimally in the control of a vehicle. This allows the classification of thoughts almost in real time with greater accuracy, in addition to giving the user the ability to make more movement decisions such as advance, stop or reverse.

References

[1] S. Sanei and J. A. Chambers. "EEG signal processing."*John Wiley and Sons*, 2013.

[2] C. Guger, A. Schlogl, C. Neuper, D. Walterspacher, T. Strein and G. Pfurtscheller. "Rapid prototyping of an eeg-based brain-computer interface (BCI)." *IEEE Transactions on Neural Systems and Rehabilitation Engineering*, 2001.

[3] A. Schlgl, D. Flotzinger and G. Pfurtschelle. "Adaptive autoregressive modeling used for single-trial eeg classification."*Biomed Technik*, 1997.

[4] S. Hayki. Adaptive Filter Theory.*Prentice-Hall*, 1986.

[5] C. Guger, G. Edlinger, W. Harkam, I. Niedermayer and G. Pfurtscheller. "How many people are able to operate an eeg-based brain-computer interface (BCI)?"*IEEE transactions on neural systems and rehabilitation engineering*, 2003.

[6] B.-G. Xu and A.-G. Song. "Pattern recognition of motor imagery eeg using wavelet transform." *Journal of Biomedical Science and Engineering*, 2008.

[7] J. L. Collinger, B. Wodlinger, J. E. Downey, W. Wang, E. C. Tyler-Kabara, D. J. Weber, A. J. McMorland, M. Velliste, M. L. Boninger and A. B. Schwartz. "High-performance neuroprosthetic control by an individual with tetraplegia." *The Lancet*, 2012.

[8] A. Astudillo, F. M. Moreno, A. Hussein and F. Garcia. "Cost-efficient brainwave controller for automated vehicles route decisions." *IEEE 20th International Conference on Intelligent Transportation Systems (ITSC)*, October 16-19, 2017, Yokohama, Japan.

[9] NeurosSky. "Thinkgear serial stream guide,"*NeurosSky*, 2015.

[10] P. Marin-Plaza, J. Beltran, A. Hussein, B. Musleh, D. Martin, A. de la Escalera and J. M. Armingol. "Stereo vision-based local occupancy

grid map for autonomous navigation in ROS."*Joint Conference on Computer Vision, Imaging and Computer Graphics Theory and Applications (VISIGRAPP)*, 2016.

[11] A. Hussein, P. Marin-Plaza, D. Martin, A. de la Escalera and J. M. Armingol. "Autonomous off-road navigation using stereo-vision and laser-rangefinder fusion for outdoor obstacles detection." *IEEE Intelligent Vehicles Symposium (IV)*, 2016.

[12] S. Ataucuri. "Ros neural."*Google Summer of Code 2015*, 2015.

[13] F. Pedregosa, G. Varoquaux, A. Gramfort, V. Michel, B. Thirion, O. Grisel, M. Blondel, P. Prettenhofer, R. Weiss, V. Dubourg, J. Vanderplas, A. Passos, D. Cournapeau, M. Brucher, M. Perrot and E. Duchesnay. "Scikit-learn: Machine learning in Python," *Journal of Machine Learning Research*, 2011.

[14] M. D. Zeiler. "ADADELTA: An Adaptive Learning Rate Method." *arXiv:1212.5701* , 2012.

[15] F. D. "Batch normalization in Neural Networks."*Towards Data Science*, 2017.

[16] E. Astrand, C. Wardak and S. B. Hamed. "Selective visual attention to drive cognitive brainmachine interfaces: from concepts to neurofeedback and rehabilitation applications." *Frontiers in Systems Neuroscience*, 2014.

4

A-RCRAFT Framework for Analysing Automation: Application to SAE J3016 Levels of Driving Automation

Elodie Bouzekri, Célia Martinie and Philippe Palanque

ICS-IRIT, Toulouse University, Toulouse, France
E-mail: elodie.bouzekri@irit.fr; célia.martinie@irit.fr; philippe.palanque.@irit.fr

Automation can be considered as a design alternative that brings the benefits of reducing the potential for human error and of increasing performance. However, badly designed automation can have unexpected consequences such as automation surprise and out-of-the-loop problems. These problems can have a very negative impact on the overall performance of the couple human/system. Theories and frameworks have been proposed to help automation designers to avoid automation issues. In this chapter, we analyse automation by decomposing it into different concepts that are emphasized separately in the literature of automation design. We present the A-RCRAFT framework that provides support for the analysis of automation design in terms of Allocation of Resources, of Control Transitions, of Responsibility, of Authority, and of Functions and Tasks (A-RCRAFT). We illustrate how this framework can be used to analyse different options of driving automation design according to the SAE J3016 levels of driving automation.

4.1 Introduction

Currently, automation is one of the main means for supporting operators or users using systems that feature increasing complexity. Automation makes it possible to reduce overall tasks complexity and effort for operators by allocating to the system tasks that were previously performed by the operator.

In aviation domain, automation makes possible to reduce the number of operators of flight crew from three to two. This reduction was possible by allocating failure detection tasks and some flying tasks to automation (e.g. the autopilot, the flight warning system or landing assistance). However, if the workload of operators is reduced when automation works correctly, in case of automation failure, extensive burden can be added to the flight crew [17]. Indeed, issues introduced by badly design automation such as complacency or out-of-the-loop problems have been identified by analysing causes of some aircraft accidents (e.g. one of cause of the crash of the Boeing 737-400 Laguardia Airport is the deactivation of the auto-throttle and the lack of awareness of the crew of this deactivation. This cause is identified in the relative National Transportation Safety Board recommendations in 1990 [33]). Then, techniques, methods and recommendations to analyse and to design interaction with automation have been proposed to deal with these issues. Among these, the "Levels of Automation" framework [38] proposes ten possible levels of automation (from level 0, where all the tasks are allocated to the human, to level 10, where all the functions are allocated to the system), as well as types of functions that can be performed by the human or by the system (information acquisition, information analysis, decision selection and action implementation) at a specified level of automation. This framework aims to provide support for the analysis of the possible allocations of tasks and functions between the operator and the system. Other types of contributions to support automation design are automation concepts such as the lumberjack analogy [43] or the automation conundrum [17] and they aim to point out the potential issues that have to be addressed when designing automation [26]. Some of them specifically aim to identify automation philosophies their implications for design and operations, in order to provide recommendations for the design of driving automation [52].

Existing approaches for the analysis of automation mainly focus on the allocation of functions and deal with authority and responsibility only at a high abstraction level. They do not provide explicit support for reasoning about the quality of a given allocation of responsibility, authority, resources and control transitions and makes the engineering of partly-autonomous system cumbersome, leaving design decisions in the hands of the programmers. We thus propose to go beyond the identification of allocation of functions and tasks and to analyse automation by identifying several other types of allocations that we gathered from the literature of automation design. We propose the A-RCRAFT framework that provides support for the analysis of automation design in terms of Allocation of Resources, of Control

Transitions, of Responsibility, of Authority, and of Functions and Tasks (A-RCRAFT). In previous work, we argued that authority and responsibility should be taken into account at design time [6]. We identified three aspects of automation that have to be identified at design time:

- what functions/tasks are allocated to the system and the human (allocation of functions and tasks),
- who can direct or prevent the execution of functions/tasks (i.e. planning) and who initiate functions/tasks (authority)
- and who is responsible for the outcome of the execution of the functions/tasks (responsibility).

The A-RCRAFT framework integrates two other aspects in addition to the three aspects presented here above: the allocation of resources and the allocation of control transition. For example, the explicit identification and description of the A-RCRAFT enables to avoid that critical tasks or functions (for which the consequences of an error or failure are catastrophic) as well as authority and responsibility on these tasks or functions are allocated to the actor (human or system) that is the less reliable for this task or function.

This chapter is structured as following. Next section (Section 2) defines the Allocation of Resources, Control transitions, Responsibility, Authority, and Functions and Tasks (A-RCRAFT) and presents the related work. Section 3 presents a qualitative analysis the Levels of Driving Automation of the SAE [22] according to the A-RCRAFT concepts previously described. We show that A-RCRAFT framework offers a conceptual background for the analysis of command and control automation and driving automation.

4.2 A Framework for Automation Analysis: A-RCRAFT

This section presents the A-RCRAFT framework for the analysis of automation in terms of Allocation of: Resources, Control transitions, Responsibility, Authority, Functions and Tasks. Each sub-section presents a concept of the A-RCRAFT framework and how these concepts are addressed in the literature. For each type of concept allocation in the framework, we first present the previous work from which the concept comes from. Then, we propose a definition illustrated by an example. We highlight the characteristics of the analysis enabled by the description of this type of concept allocation. We conclude each section by the identification of the output of the description of the corresponding type of allocation. In the first sub-section, we define the concept of Allocation of Functions and Tasks (**A**-RCRA**FT**). This sub-section is more detailed than the other ones because it introduces primary concepts that are needed for the decomposition of allocation of resources, control transition, responsibility and authority.

4.2.1 Allocation of Functions and Tasks

Existing approaches dealing with automation analysis usually focus on identifying tasks and functions that should be allocated to either the operator or the system. These approaches provide support for the identification of which functions are good candidate to be performed by the system and of which tasks should be performed by the operator [7, 13, 15] and [50].

We define Allocation of Functions and Tasks as the identification of the functions the system performs and of the tasks the human performs. Human tasks are composed of the set of: perceptive, cognitive, motor and input interactive tasks that the human performs. System functions are composed of the set of: sense, analysis, selection and output functions that the system performs.

For example, in civil aircraft cockpit, the pilot flying performs perceptive, cognitive, motoric and interactive input tasks relative to fly the aircraft. The pilot monitoring performs perceptive, cognitive, motoric and interactive input tasks relative to manage systems. The flight warning system (FWS) performs failure detection, analysis, selection of possible options for solving the problem and display options relative to failure management.

Furthermore, we complete this definition by adding:

- the identification of the temporal ordering of the system functions, of the human tasks and of the temporal ordering between system functions and human tasks [28] [29].
- the identification of resources required by the human and by the system to perform their functions and tasks [30], which are the devices, data and objects.

Information processing decomposition of a function or a task

Information processing can be decomposed in several steps, whether they are performed by the human or by the system. The Action Theory [35], the Human Model Processor [10] or the SRK model [39] decompose human information processing in several steps in order to support the analysis of the different types of possible user actions (e.g. perceptive or cognitive actions for the Human Model Processor). In an analogous way, several contributions to automation analysis propose to identify different types of functions. Dearden

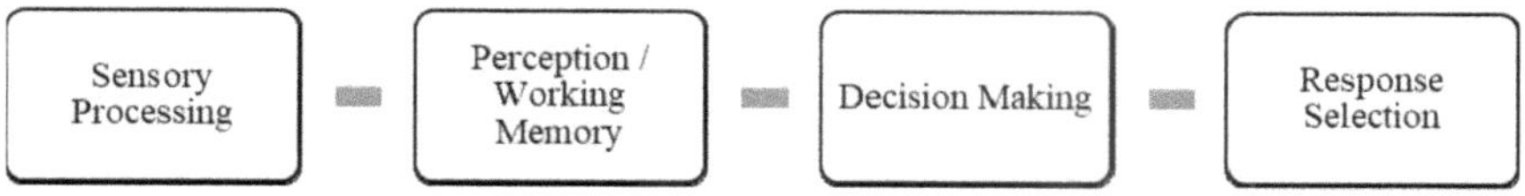

Figure 4.1 Simple four-stage model of human information processing from [38].

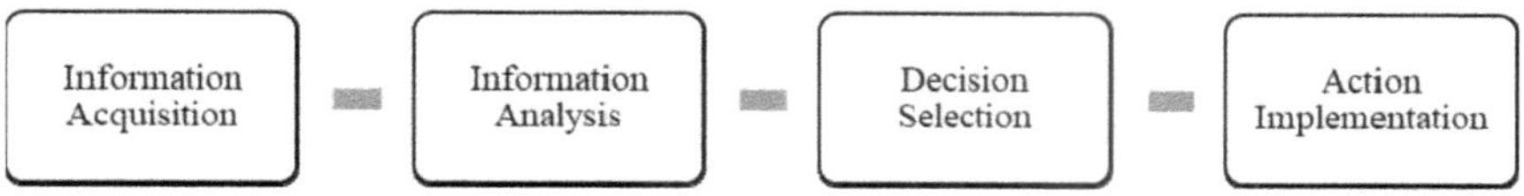

Figure 4.2 Simple four-stage model of system information processing from [38].

et al. [12] propose four categories of functions: information, decision, action and supervision. Kaber and Endsley [24] propose four generic types of functions: monitoring (e.g., scanning visual displays), generating (producing options or task strategies), selecting (choosing a particular option or strategy) and implementing (carrying out the chosen option). Parasuraman et al. [38] propose the equivalent for a system (see Figure 4.2) of a simplified four stages model of human information processing (see Figure 4.1) to decompose a function: information acquisition, information analysis, decision selection and action implementation. We use the information processing decompositions of Parasuraman et al. [38] in the remainder of the chapter.

Analysis enabled by the identification of the Allocation of Functions and Tasks

The description of the allocation of functions and tasks is necessary to identify the optimal distribution of both functions and tasks between a partly-autonomous system and a human.

The allocation of functions and tasks is also central to the automation analysis because it provides support to migrate human tasks to be performed by the system or to migrate system functions to be performed by the human [28] [29]. More precisely, the allocation of functions and tasks enables to describe functions that replace or enhance (e.g. a radar system performs a function that humans are not capable to perform) or support partially or entirely tasks performed by humans. Indeed, according to [51], not enough functions allocated to human will lead to underload and boredom and thus decreased performance while too many functions will lead to cognitive, perceptive or motoric overload and increase stress and likelihood of user errors.

The output of the allocation of functions and tasks is:

- the description of the temporally ordered sets of functions that the system should perform to support user goal
- the description of the temporally ordered sets of tasks that the user should perform to reach her/his goal.
- The description of the temporal ordering between system functions and user tasks, as well as of the data, objects and devices manipulated by both the system and the user.

This implies that, during automation analysis, all the system functions and all the human tasks have to be identified with the appropriate level of abstraction in order to be able to describe the allocation of tasks and functions. Such description can have the form of a task model as shown in [28–30].

Such descriptions enable to analyse automation according to the following secondary concepts: resources (e.g. data, objects and devices required to perform a function or task), control transition (e.g. temporal ordering between system function and human task), responsibility (e.g. the influence of system or human action on a resource) and authority (e.g. description of who can trigger and/or delegate a tasks or function). These secondary concepts are detailed in their respective section hereafter.

4.2.2 Allocation of Resources

In the same way that Norman emphasizes the problem of communication feedback and feedforward with automation through civil aircraft examples [36], Bradshaw et al. [9] alert about the potential issues of low observability and low understandability caused by highly automated but silent systems. Understandability is the ability to "form a mental model and predict future system behaviour" [45]. Mental models are constructed from "the system image" [35] (i.e. system perceivable outputs, appearance and documentation). Explication of intention is the ability of the system to "display or says that it will act in a particular way" [45]. Battiste et al. [4] highlights the need of transparency for a good teamwork with automation. Jansson et al. [23] define GMOC model for the design of automation in which observability enhances the construction of mental model and the formulation of goals. Woods highlights the importance to deal with the observability (interfaces and feedback issues) problem when interacting with automated entity to avoid automation surprise [49]. Starter and Woods [41] define observability as "*the ability of available feedback to actively support operators in monitoring and staying ahead of system activities and transitions*". They precise that observability is not only information availability but depends on the cognitive work needed to analyse information. More precisely than observability, the term "transparency" is the extent to which a system allows the operator to perceive and understand the information required for her/his activities and to perceive and understand the behaviour of the system [4, 5]. Some existing levels of automation integrate the concept of allocation of Information. For example, the Levels of Automation of Parasuraman [38] indicate what type of information is presented to the human and the frequency of information exchanges. Some studies highlight the need of a shared understanding through effective information exchanges from the system to the human and from the human to the system. For example, Klien et al. [25] propose to exchange information to maintain shared knowledge, goals and intentions between the human and the system.

We extend the concept of allocation of Information and define Allocation of Resources as the identification of the data, devices and objects required to perform the system functions and the human tasks, which are exchanged (or not) between the human(s) and system(s). For example, in civil aircraft cockpits, active alarms, on-going processes and system data are software objects in the system that are then provided by the system on output devices, from which the pilot acquires information. In specific contexts, some of the on-going alarms are not displayed to the user (e.g. while take-off). Data provided to the human by the system can vary in term of frequency and in term of type.

Analysis supported by the identification of the allocation of resource

The description of the allocation of Resource provides support to analyse the transparency of system activities. For example, if the human cannot perceive any information from the system for a part of the activity or for its behaviour, the system is not transparent for that part of the activity or for its behaviour.

The description of the allocation of Resources also provides support to analyse the understandability and explication of intention of system activities. These properties are identified as trust attributes [45]. The description of types of information exchanged enable understandability analysis [35]. Types of information provided by the system enable the analysis of intention of system property. Transparency and understandability properties are not restricted to automation analysis but become critical as they enable awareness which is necessary when human operates with automation [17].

The output of the identification of allocation of Resources is the description of provided resources (data, objects and devices), of the parts of the system concerned by these exchanges or the human concerned by these exchanges, and of the resources not released by each part of the system or by the human.

4.2.3 Allocation of Control Transitions

Lu et al. [27] propose a classification tree of control transitions based on research in transitions of control for driving systems. Other studies propose taxonomies of transitions of control for driving automation [47, 31, 48]. As identified in the aviation domain (e.g. [42, 17]), control transitions can be an issue during automation design as they require situation awareness, no mental overload and avoiding out-of-the-loop problem. Adaptive automation has been proposed to deal with human workload regulation and out-of-the-loop problem [44]. The adaptive automation allows changes of the allocation of functions and tasks depending on the environment or on the human conditions [44]. Control transition taxonomies of [31] integrate the scheduling (scheduled, not scheduled) and [47] integrates time allowed to takeover (immediate and not). Loer et al. [24] propose a model-checking technique to verify the relevance of all possible temporal scheduling dynamic allocation of function (transitions of control). [7, 13, 15] and [50] approaches provide support for describing the possible transitions of control between the human and the system. We propose a definition to allow the description of such control transitions, regardless the field of automation analysis (aviation or driving automation), to be able to describe control transition whatever its implementation.

We define Allocation of Control Transitions as the identification of changes from an allocation of functions and tasks to another that can occur under temporal constraint during the activity. For example, in civil aircraft cockpit, while the pilot monitoring controls the tasks relative to manage aircraft systems, the flight warning system can trigger an alarm that the pilot monitoring must handle. Then, a control transition occurs. The pilot monitoring on-going task is interrupted and she has to manage the alarm. Another example in an automated car is that the system drives the car while the human supervises the running car. Then, the system may ask the human to resume driving within 2 minutes due to a traffic situation that the system cannot handle. A control transition occurs: the human switches from supervising task to driving task and the system no longer controls the driving task. Control transition may also occur partially for a sub task or step by step to lead to the full transition. For example, the car can hand over driving step by step: first it takes the longitudinal control, then the lateral control [47]. We can also imagine cases where the human take over only the longitudinal control leaving the lateral control to the car (i.e. the cruise control). In that case, the human and the car share the driving task and the control transition occurred partially. The allocation of control transitions must be done with a conscientious analysis of tasks, functions and resources humans and systems exchange as control transitions may require the ability to understand and respond to events. This is particularly the case for a type of control transition where the system asks the human to resume a task like handover driving task in autonomous or partly autonomous cars [31]. This example highlights that control transitions can have temporal constraints. A control transition can be scheduled (i.e. from that defined moment, human and system controls other tasks and functions). The control transition can also have limited time on the change of allocation of functions and tasks. For example, human is asked to validate or not a proposition within a limited time (veto).

Analysis supported by the description of the allocation of control transitions

The description of changes from an allocation of functions and tasks to another provides support for the identification of possible intervention of the human (or the system) during the task (or function) of the system (or of the human). It also provides support for the identification of possible interruption of the task (or function) of the human (or of the system) by the system (or by the human). These identifications can be useful to prove the compliance of a system design with certification specifications. In the case where the automation should enable the crew to intervene manually in any function [16], the description of the allocation of control transition can enable to verify that there is at least one possible control transition for each system functions.

The output of control transition allocation is the description of each possible changes of tasks and functions allocation and of their temporal constraints.

4.2.4 Allocation of Responsibility

Whereas Responsibility is considered as a concept that has to be taken into account at design time, the contributions that we have found do not explicitly support the systematic analysis of allocation of responsibility. Flemisch et al. [18] as well as Miller and Parasuraman [32] proposed a conceptual framework that highlight the importance of analysing responsibility at design time.

However, this framework for analysing responsibility is abstract which makes it not useful for describing the activity at the task level.

We define Allocation of Responsibility as the identification of who can cause a derivation of the expected result of the activity. The allocation of responsibilities (between human and system) must make explicit the outcomes that are relevant, and who (the user or the system) influences these outcomes, in order to identify what actor should be accountable for the result of an action [18]. For example, in civil aircraft cockpits, the pilot monitoring has the responsibility to judge correctly the pertinence of alarms (spurious or not). The wrong management of alarms may cause a derivation of the expected result of the fly which is to travel safely from airport A to airport B.

Analysis supported by the description of the allocation of responsibility

The analysis of the description of the allocation of responsibility provides support for the identification of the actor who has been at the root cause of an unwanted or unexpected outcome [6].

The output of the allocation of responsibilities consists in a list of both all expected and actual outcomes when an activity is performed and identification of functions and tasks that can cause a deviation of these outcomes. The comparison between actual outcomes and expected outcomes makes it possible to identify these deviations (that could be errors on the user side or failures on the system side).

4.2.5 Allocation of Authority

Whereas Authority is considered as a concept that has to be analysed at design time, the contributions that we have found do not explicitly support the systematic analysis of allocation of authority. Flemisch et al. [18] as well as Miller and Parasuraman [32] proposed a conceptual framework that highlight the importance of analysing authority at design time. Boy [8] proposed a conceptual model to support the analysis of authority sharing amongst several humans and systems. Gombolay et al. [21] considers that the identification of "the right to make decisions" has to be done explicitly through allocation of authority. Cummings and Bruni [11] proposed to extend Parasuraman information processing model by adding a decision making component. However, frameworks for analysing authority are abstract which makes them not useful for describing the activity at the task level.

We define Allocation of Authority as the identification of who can initiate tasks or functions and who plans future tasks or functions that the human(s) or system(s) will perform. For example, in civil aircraft cockpits, the pilot monitoring can cancel an alarm if she judges it spurious. The pilot monitoring expresses her authority by planning the system next task on suppressing alarm. Another example in automated cars is that the system can initiate a control transition of the driving task if it judges that the human drives hazardously. The system expresses its authority by initiating a Control Transition (defined in sub-section 2.1.3) and planning the human next task on supervision.

Analysis supported by the description of the allocation of authority

The description of the allocation of authority provides support for the identification of the allocation of control transition. It makes it possible to identify who can intervene in or interrupt the tasks (or functions) of the human or of the system. The description of who can initiate or plan a task in addition to the description of who can plan the tasks of the other entities (e.g. give orders) and the allocation of tasks and functions provides support for the definition of the relationship between the entities. For example, if the system selects actions to perform (i.e. authority by planning future actions) and implement them, and that the human can intervene during the execution of these system functions (i.e. initiate a control transition), the relationship between the entities is named "supervisory control" as defined in [17].

The output of the identification of the allocation of authority is the description of what the system(s) and the human(s) can plan (e.g. orders or decisions), and what functions or tasks they can initiate.

4.3 Qualitative Analysis of SAE J3016 Levels of Driving Automation with A-RCRAFT

This section presents the results of the qualitative analysis of the SAE J3016 Recommended Practice "*describes motor vehicle driving automation systems that perform part or all of the dynamic driving task (DDT) on a sustained basis*" [22]. It also defines a taxonomy of six levels of automation ranging from no automation to full automation of the driving task. In this section, we present the results from the qualitative analysis of this taxonomy according to the allocation of resources, control transitions, responsibility, authority, and allocation of functions and tasks (A-RCRAFT). We first present the main target user tasks and system functions that are the focus of the SAE J3016 recommendations. We then decompose each level of automation according to the A-RCRAFT framework. At last, we highlight the issues that the A-RCRAFT framework enables to point out.

4.3.1 Scope of the SAE J3016 for the Human Tasks and System Functions

The main tasks and functions for which the allocation between the human and the system is specified in the SAE J3016 Recommended Practice [22] are gathered under the term of Dynamic Driving Task (DDT). This term gathers the following actions which, depending on the level of automation, are performed by the human or by the system:

- Lateral vehicle motion control via steering (operational),
- Longitudinal vehicle motion control via acceleration and deceleration (operational),
- Monitoring the driving environment via object and event detection, recognition, classification, and response preparation (operational and tactical),
- Object and event response execution (operational and tactical),
- Manoeuvre planning (tactical),
- Enhancing conspicuity via lighting, signalling and gesturing, etc. (tactical).

Crosswise to these tasks/functions, the SAE J3016 specifies a set of tasks/function under the acronym OEDR that stands for Object and Event Detection and Response.

In addition, the SAE J3016 explicitly label the part of the system that is concerned to perform the driving tasks. It is named ADS (Automated Driving System). And the SAE J3016 also explicitly recognize that a driving task may be limited to an Operational Design Domain (ODD). Examples of ODD are: fast lane, highway...

4.3.2 Decomposition of Levels of Driving Automation According to A-RCRAFT

Table 4.1 presents the decomposition of the SAE J3016 levels of automations [22] according to the A-RCRAFT framework. The first and second columns contain the Levels of Driving Automation (LDA) as defined in the SAE J3016, ranging from the lowest automation (Level 0) called "no driving automation" to the highest automation (Level 5) called "full driving automation". The other columns describe the characteristics of each level of automation, from the information available in the SAE J3016, according to the A-RCRAFT concepts.

4.3.3 Results of the Analysis and Benefits from Using A-RCRAFT

The decomposition of the SAE J3016 levels of automation according to the A-RCRAFT framework enables to explicitly point out ambiguities and missing data that may be confusing during automation analysis:

- The decomposition according to the Resource concept of A-RCRAFT (column "Resources" in Table 1) enables to see that across all levels there are very few descriptions about the information manipulated on the human side, about the data and devices on the system side and about the exchange of information and data between the human and the system. For levels 1 and 2, there is no description of the information that the user should have access. For these levels, there is no description of the data that the system should have access to. For levels 3, 4 and 5, two types of information that should be available to the user are described: operational readiness of the automation, and request to intervene from the system. For level 3, additional information should be made available to the user and is information about performance-relevant system failures.
- The decomposition according to the Control Transition concept of A-RCRAFT (column "Control Transition" in Table 1) enables to see that possible control transitions are explicitly stated for the tasks of lateral and longitudinal vehicle motion (for levels 1 and 2), as well as for the entire DDT (for levels 3,4 and 5). There are explicit qualitative recommendations about the possible time frames for the transitions only for the disengagement of automation at levels 3, 4 and 5. At level 3, the transition from the system to the user should be immediate whereas there may be a delay at levels 4 and 5.
- The decomposition according to the Responsibility concept of A-RCRAFT (column "Responsibility" in Table 1) enables to see that even if automation levels are high, the human still has responsibilities. This decomposition also enables to highlight that between level 3 and 4, the human has the same responsibilities whereas the human has more

Table 4.1 Levels of Driving Automation from [22] and its interpretation using A-RCRAFT

LDA	Definition	Resources	Control Transitions	Responsibility	Authority	Functions and Tasks
0 No Driving Automation	The performance by the driver of the entire DDT, even when enhanced by active safety systems.	No type and frequency specified	No control transitions	All to the human	All to the human	Human: entire DDT. System: active safety mechanisms (e.g. electronic stability control, automated emergency braking)
1 Driver Assistance	The sustained and ODD-specific execution by a driving automation system of either the lateral or the longitudinal vehicle motion control subtask of the DDT (but not both simultaneously) with the expectation that the driver performs the remainder of the DDT.	No type and frequency specified	Control transition of lateral motion or longitudinal motion to the system or from the system to the human	All to the human	All to the human	Human: entire DDT except lateral **or** longitudinal vehicle motion if allocated to the system, supervision of driving automation when engaged. System: Lateral vehicle motion **or** Longitudinal vehicle motion if automation engaged.
2 Partial Driving Automation	The sustained and ODD-specific execution by a driving automation system of both the lateral and longitudinal vehicle motion control subtasks of the DDT with the expectation that the driver completes the OEDR subtask and supervises the driving automation system.	No type and frequency specified	Control transition of both lateral motion and longitudinal motion from the system to the user	All to the human.	All to the human	Human: entire DDT except lateral **and** longitudinal vehicle motion if allocated to the system, supervision of driving automation when engaged. System: Lateral and longitudinal motion if automation engaged.

3 Conditional Driving Automation	The sustained and ODD-specific performance by an ADS of the entire DDT with the expectation that the DDT fallback-ready user is receptive to ADS-issued requests to intervene, as well as to DDT performance-relevant system failures in other vehicle systems, and will respond appropriately.	Information about the operational readiness of the automation, request to intervene from the system, performance relevant system failures	Control transition of the entire DDT from the user to the system or from the system to the user immediately after request from the user.	To the system for enabling the engagement for an ODD, and if is engaged to monitor performance-relevant system failures, to analyse how to achieve a minimal risk condition and to request the human to intervene. To the human verify the operational readiness of the automation, to engage it and if engaged to monitor performance-relevant system failures, to analyse how to achieve a minimal risk condition and to request disengagement if necessary.	To the human for triggering the engagement for an ODD and for triggering a fall back. To the system for enabling the engagement for and ODD an for requesting the human to intervene.	Human: entire DDT if automation is disengaged. If automation is engaged user has to wait for a request to intervene, to monitor performance-relevant system failures, to analyse how to achieve a minimal risk condition and to request to disengage if relevant. System: entire DDT if automation is engaged, monitor and analyse performance-relevant system failures, request the driver to intervene, disengage an appropriate time after issuing a request to intervene or upon driver request.

(Continued)

Table 4.1 Continued

LDA	Definition	Resources	Control Transitions	Responsibility	Authority	Functions and Tasks
4 High Driving Automation	The sustained and ODD-specific performance by an ADS of the entire DDT and DDT fall back without any expectation that a user will respond to a request to intervene.	Information about operational readiness of the automation, request to intervene from the system.	Control transition of the entire DDT from the user to the system or from the system to the user but there may a delay for the transition from the system to the user.	To the system if automation for enabling the engagement for an ODD and if is engaged. To the human verify the operational readiness of the automation and to engage it	To the human for triggering the engagement for an ODD, for requesting a fall back and for accepting a request to intervene. To the system for enabling the engagement for an ODD, for requesting the human to intervene and for accepting a fall back.	Human: if automation is engaged: verify the operational readiness, engage automation, optionally perceive and/or take into account a request to intervene, optionally request to disengage and to achieve a minimal risk condition. Entire DDT if automation is disengaged. System: allow the engagement for an ODD, then if automation is engaged: entire DDT, monitor and analyse performance-relevant system failures, request the driver to intervene, disengage if user answers to the request or fall back to a minimal risk condition and disengage if the user did not answer, disengage if the user performs the DDT.

5 Full Driving Automation	The sustained and unconditional (i.e., not ODD-specific) performance by an ADS of the entire DDT and DDT fall back without any expectation that a user will respond to a request to intervene.	Information about operational readiness of the automation, request to intervene from the system.	Control transition of the entire DDT from the user to the system or from the system to the user but there may a delay for the transition from the system to the user.	To the system if automation for enabling the engagement and if is engaged To the human verify the operational readiness of the automation and to engage it	To the human for triggering the engagement, for requesting a fall back and for accepting a request to intervene. To the system for enabling the engagement, for requesting the human to intervene and for accepting a fall back.	Human: if automation is engaged: verify the operational readiness, engage automation, optionally perceive and/or take into account a request to intervene, optionally request to disengage and to achieve a minimal risk condition. Entire DDT if automation is disengaged. System: allow the engagement for all driver manageable on-road conditions, then if automation is engaged: entire DDT, monitor and analyse performance-relevant system failures, request the driver to intervene, disengage if user answers to the request or fall back to a minimal risk condition and disengage if the user did not answer, disengage if the user performs the DDT.

information about the system state at level 3 than at level 4 (information about performance relevant system failures at level 3). Having these additional information, the human may better fulfil her responsibilities. This is maybe implicitly taken into account in the recommendations because for level 3 the system has to disengage immediately after human requested (meaning that the human will for sure make no errors?) whereas at level 4, the system may delay the human request (meaning that the human could have taken a wrong decision as s/he does not have a relevant mental model of the system current state?). In addition, this decomposition also enables to highlight that the system and the human have the same responsibilities at level 3. They both are responsible for monitoring performance-relevant system failures and to act if necessary (ask for disengagement from the human and ask for human intervention on the system side).

- The decomposition according to the Authority concept of A-RCRAFT (column "Authority" in Table 1) enables to see that the system has higher authority at levels 4 and 5, which seems compliant with the fact that it is getting more autonomous. At level 3, once the human requested to disengage automation, the system has to give back immediately the DDT task whereas for levels 4 and 5, the system can decide or not to give back the DDT task to the human. For these two upper levels, it is also interesting to note that even if the system is getting more authority, it cannot force the human to take back the DDT task (user has authority for accepting or not a fall back).
- The decomposition according to the Functions and Tasks concept of A-RCRAFT (column "Functions and Tasks" in Table 1) enables to explicitly describe the migration of the tasks from the human to the system. From level 0 to level 2, the lateral and longitudinal motion control is first assigned to the human and is then progressively assigned to the system. From level 3 to 5, it is interesting to note that there is few migration as the entire DDT task can be assigned to the system and to the human. The few differences in terms of allocation of functions and tasks concerns the additional task of monitoring performance-relevant failures for the human at level 3, and the scope of the automation engagement between levels 3, 4 and level 5 (ODD specific for levels 3, 4 and all driving conditions for level 5).

Each of the concept is important for the analysis. To analyse the allocation of functions and tasks is not enough to compare different levels of automation. For an equivalent allocation of functions and tasks between levels of automation, the allocation of resources, control transition, responsibility and authority can be different. For example, we have shown that the decomposition according to the A-RCRAFT framework enables to highlight that even if the differences in terms of allocation of functions and tasks are not high between levels 3, 4 and 5, there are big differences in terms of allocation of resources, control transition, responsibility and authority, as detailed in the previous paragraphs (e.g. the authority on the fall back between level 3 and level 4 and the information available to the user between level 3 and level 4).

4.4 Conclusion

Automation has been studied for many years and even though metaphors [19] or frameworks [38, 12] have been proposed, the description of the allocation of resources, control transition, authority, responsibility, and functions and tasks between the human and the system is not explicitly supported together in a single framework. This paper has introduced such a framework called A-RCRAFT. It allows to systematically identify and describe the Allocation of Function and Tasks, together with the Allocation of Resources, Control Transition, Authority and Responsibility.

We have shown how A-RCRAFT framework can be fruitfully used to analyse the SAE J3016 standard. In this context, this analysis has shown that: (i) the allocation of resources between system and operator is poorly described, (ii) there are partial description of the timeframe for control transitions (only for the disengagement of automation), (iii) the human has the same responsibilities between level 3 and level 4 whereas s/he is being provided with less information at level 3, (iv) at the highest level of automation the human still has authority for specific tasks, v) there is a progressive migration of the driving tasks from the lowest levels to the highest levels.

We have also applied the A-RCRAFT framework to the Parasuraman et al. LoA [38] (not presented here due to space constraints) demonstrating its ability to analyse automation aspects in various domains.

Even though we have demonstrated the utility of A-RCRAFT for analysing levels of automation descriptions, it can be also used for automation design of partly-autonomous interactive systems. Throughout the design process, allocating the five dimensions of RCRAFT must be carefully considered in order to ensure (i) the allocation of function or task to the best player, (ii) the allocation of the resources to support adequate workload, (iii) the careful design of control transitions especially in case of unexpected events,

(iv) the understanding of authority sharing between the system and the operator (especially by means of exploitation of control transitions), (v) the understanding of where the responsibilities lay to comply with regulations and law.

To support these design activities processes, techniques and tools are needed. We have extended task modelling notations to support the identification and representation of authority and responsibility along with the description of allocation of functions and tasks as well as of allocation of resources [6]. We plan to work on a tool-supported process to support the systematic identification and representation of the five dimensions of the A-RCRAFT framework during automation design.

References

[1] Airbus A350 Flight Crew Operating Manual, 5T1 A350 FLEET FCOM. Technical Report. Airbus.

[2] Barboni, E., Ladry, J-F., Navarre, D., Palanque, P., Winckler, M. (2010) "Beyond Modelling: An Integrated Environment Supporting Co-Execution of Tasks and Systems Models," in Proc. of EICS '10. ACM, 143–152.

[3] Basnyat, S., Navarre, D., Palanque, P. (2008). "Usability Service Continuation through Reconfiguration of Input and Output Devices in Safety Critical Interactive Systems," in Proceedings of International Conference on Computer Safety, Reliability and Security (SAFECOMP 2008), Newscastle, UK, Vol. LNCS, Springer-Verlag,

[4] Battiste, V., Lachter, J., Brandt, S., Alvarez, A., Strybel. T.Z., and Vu, K-PL. (2018). "Human-Automation Teaming: Lessons Learned and Future Directions." Human Interface and the Management of Information. Information in Applications and Services, Springer International Publishing, 479–493.

[5] Bernhaupt, R., Cronel, M., Manciet, F. Martinie, C. and Palanque P.. 2015. Transparent Automation for Assessing and Designing better Interactions between Operators and Partly-Autonomous Interactive Systems. In Proceedings of the 5th International Conference on Application and Theory of Automation in Command and Control Systems (ATACCS '15). Association for Computing Machinery, New York, NY, USA, 129–139.

[6] Bouzekri E, Canny A, Martinie C, Palanque P, and Gris C. Using Task Descriptions with Explicit Representation of Allocation of Functions,

Authority and Responsibility to Design and Assess Automation. 2019. In Human Work Interaction Design. Designing Engaging Automation (IFIP Advances in Information and Communication Technology), 36–56.

[7] Boy, G. Cognitive Function Analysis for Human-Centered Automation of Safety-Critical Systems. Proceedings of ACM CHI 1998, 265–272 (1998).

[8] Boy, G. Orchestrating Situation Awareness and Authority in Complex Socio-technical Systems. CSDM 2012: 285–296

[9] Bradshaw J. M., Hoffman R. R, Woods D. D., and Johnson M. The Seven Deadly Myths of "Autonomous Systems." IEEE Intelligent Systems 28, 3: 54–61 (2013).

[10] Card S, Moran T, and Newell A. The model human processor- An engineering model of human performance. 1986. Handbook of perception and human performance. 2, 45–1.

[11] Cummings M. L. and Bruni S. 2009. Collaborative Human– Automation Decision Making. In Springer Handbook of Automation (pp. 437-447). Springer Berlin Heidelberg. LNCS Homepage, http://www.springer.com/lncs, last accessed 2016/11/21.

[12] Dearden A. IDA-S: A Conceptual Framework for Partial Automation. 2001. In People and Computers XV—Interaction without Frontiers, 213–228. https://doi.org/10.1007/978-1-4471-0353-0_13

[13] Dearden, A., Harrison, M. D., Wright, P.C. Allocation of function: scenarios, context and the economics of effort. Int. J. Hum.-Comput. Stud. 52(2): 289–318 (2000).

[14] Dictionary. English dictionary. www.dictionary.com/browse/automation, last accessed September 2018.

[15] Dittmar, A., Forbrig, P. Selective modeling to support task migratability of interactive artifacts. In Proc. of the 13th IFIP TC 13 international conference on Human-computer interaction - Volume Part III (INTERACT'11), Vol. Part III. Springer-Verlag, Berlin, Heidelberg, 571–588 (2011).

[16] EASA. Cs-25 – Certification Specification and Acceptable Means of Compliance for Large Aeroplanes. 2015.

[17] Endsley MR. From Here to Autonomy: Lessons Learned From Human–Automation Research. 2017. Human Factors 59, 1: 5–27. https://doi.org/10.1177/0018720816681350

[18] Flemisch, F., Heesen, M., Hesse, T., Kelsch, J., Schieben, A., Beller, J. Towards a dynamic balance between humans and automation: Authority,

ability, responsibility and control in shared and cooperative control situations. Cognition, Technology & Work, 14(1) (2012), pp. 3–18.

[19] Flemisch F., Adams C., Conway S., Goodrich K. et al. The H metaphor as a guideline for vehicle automation and interaction, 1975, NASA TM, 2003-212672.

[20] Fahssi R., Martinie C., Palanque P: Enhanced Task Modelling for Systematic Identification and Explicit Representation of Human Errors. INTERACT (4) 2015: 192–212.

[21] Gombolay, M. C., Gutierrez, R. A., Clarke, S. G., Sturla, G. F. and Shah, J. A. 2015. Decision-making authority, team efficiency and human worker satisfaction in mixed human—robot teams. Auton. Robots 39, 3 (October 2015), 293–312.

[22] J3016 Taxonomy and Definitions for Terms Related to On-Road Motor Vehicle Automated Driving Systems SAE International: (2014).

[23] Jansson A, Stensson P, Bodin I, Axelsson A and Tschirner S. Authority and Level of Automation. 2014. Human-Computer Interaction. Applications and Services, Springer, Cham, 413–424.

[24] Kaber DB and Mica R. Endsley. The effects of level of automation and adaptive automation on human performance, situation awareness and workload in a dynamic control task. 2004. Theoretical Issues in Ergonomics Science 5, 2: 113–153. https://doi.org/10.1080/146392 2021000054335

[25] Klien G, Woods DD, Bradshaw JM, Hoffman RR, and Feltovich PJ. Ten challenges for making automation a "team player" in joint human-agent activity. 2004. IEEE Intelligent Systems 19, 6: 91–95.

[26] Lee JD and Seppelt BD. Human Factors in Automation Design. 2009. In S.Y. Nof, ed., Springer Handbook of Automation. Springer Berlin Heidelberg, Berlin, Heidelberg, 417–436.

[27] Lu Z, Happee R, Cabrall CDD, Kyriakidis M, and De Winter JCF. Human factors of transitions in automated driving: A general framework and literature survey. 2016. Transportation Research Part F: Traffic Psychology and Behaviour 43: 183–198.

[28] Martinie C., Palanque P., Barboni E., Ragosta M. Task-model based assessment of automation levels: Application to space ground segments. IEEE Conference on System, Man and Cybernetics, 2011, pp. 3267–3273.

[29] Martinie C., Palanque P., Barboni E., Winckler M., Ragosta M., Pasquini A., Lanzi P. Formal Tasks and Systems Models as a Tool for Specifying and Assessing Automation Designs. 1st international Conference

on Application and Theory of Automation in Command and Control Systems, (ATACCS 2011) Barcelona, Spain, May 2011, ACM DL.

[30] Martinie, C., Palanque, P., Bouzekri, B., Cockburn, A., Canny, A., Barboni, E. 2019. Analysing and Demonstrating Tool Supported Customizable Task Notations. In Proceedings of the ACM on Human-Computer Interaction, Vol. 3, EICS, Article 1, 2 June 2019), 26 pages.

[31] McCall R, McGee F, Mirnig A, et al. A taxonomy of autonomous vehicle handover situations. 2019. Transportation Research Part A: Policy and Practice 124: 507–522.

[32] Miller, C., A., Parasuraman, R. Designing for flexible interaction between humans and automation: delegation interfaces for supervisory control. Human Factors, 49, 57–75 (2007).

[33] National Transportation Safety Board. Aircraft accidents report: USAIR, Inc., Boeing 737-400, LaGuardia Airport, Flushing New York, September 20, 1989 (NTSB/AAR-90-03). 1990.

[34] NHTSA. https://www.transportation.gov/transition/nhtsa-top-policy-issues. 2017.

[35] Norman D. The design of everyday things: Revised and expanded edition. 2013. Basic books.

[36] Norman DA. The 'problem' with automation: inappropriate feedback and interaction, not 'over-automation.' 1990. Philosophical Transactions of the Royal Society of London. B, Biological Sciences 327, 1241: 585–593.

[37] Oxford. English Dictionnary. https://en.oxforddictionaries.com/definition, last accessed April 2018.

[38] Parasuraman, R.; Sheridan, T.B. & Wickens, C.D. "A model for types and levels of human interaction with automation" Systems, Man and Cybernetics, Part A: Systems and Humans, IEEE Trans. on, vol. 30, no. 3, pp. 286–297, May 2000.

[39] Rasmussen J. Skills, rules, and knowledge; signals, signs, and symbols, and other distinctions in human performance models. 1983. IEEE Transactions on Systems, Man, and Cybernetics SMC-13, 3: 257–266.

[40] RTCA. Do-178c – Software Considerations in Airborne Systems and Equipment Certification. 2012.

[41] Nadine B. Sarter and David D. Woods. 1997. Team Play with a Powerful and Independent Agent: Operational Experiences and Automation Surprises on the Airbus A-320. Human Factors 39, 4: 553–569. https://doi.org/10.1518/001872097778667997

[42] Sarter NB, Woods DD, and Billings CE. Automation surprises. 1997. Handbook of human factors and ergonomics 2: 1926–1943.

[43] Sebok A and Wickens CD. Implementing Lumberjacks and Black Swans Into Model-Based Tools to Support Human–Automation Interaction. 2017. Human Factors 59, 2: 189–203. https://doi.org/10.1177/0018720816665201

[44] Sheridan T. B. Adaptive Automation, Level of Automation, Allocation Authority, Supervisory Control, and Adaptive Control: Distinctions and Modes of Adaptation. 2011. IEEE Transactions on Systems, Man, and Cybernetics - Part A: Systems and Humans 41, 4: 662–667. https://doi.org/10.1109/TSMCA.2010.2093888

[45] T. B. Sheridan. 1988. Trustworthiness of Command and Control Systems. IFAC Proceedings Volumes 21, 5: 427–431. https://doi.org/10.1016/S1474-6670(17)53945-2

[46] Vagia M., Transeth A. A., Fjerdingen S. A. A literature review on the levels of automation during the years. What are the different taxonomies that have been proposed? Applied Ergonomics 53: 190–202 (2016).

[47] Walch M, Lange K, Baumann M, and Weber M. Autonomous Driving: Investigating the Feasibility of Car-driver Handover Assistance. 2015. In Proceedings of the 7th International Conference on Automotive User Interfaces and Interactive Vehicular Applications (AutomotiveUI '15), 11–18. https://doi.org/10.1145/2799250.2799268

[48] Wintersberger P, Green P, and Riener A. Am I Driving or Are You or Are We Both? A Taxonomy for Handover and Handback in Automated Driving. 2017. Proceedings of the 9th International Driving Symposium on Human Factors in Driver Assessment, Training, and Vehicle Design: driving assessment 2017, University of Iowa, 333–339.

[49] Woods DD. Decomposing Automation: Apparent Simplicity, Real Complexity. 26.

[50] Wright, P. C., Dearden, A., Fields, B. Function allocation: a perspective from studies of work practice. Int. J. Hum.-Comput. Stud. 52(2): 335–355 (2000)

[51] Yerkes RM, Dodson JD (1908). "The relation of strength of stimulus to rapidity of habit-formation". Journal of Comparative Neurology and Psychology 18: 459–482.

[52] Young MS, Stanton NA, and Harris D. Driving automation: learning from aviation about design philosophies. 2007. International Journal of Vehicle Design 45, 3: 323. https://doi.org/10.1504/IJVD.2007.014908

5

Autonomous Vehicles: Vulnerable Road User Response to Visual Information Using an Analysis Framework for Shared Spaces

Walter Morales Alvarez and Cristina Olaverri-Monreal

Johannes Kepler University Linz, Austria;
Chair Sustainable Transport Logistics 4.0
E-mail: walter.morales_alvarez@jku.at; cristina.olaverri-monreal@jku.at

Autonomous vehicles (AV) have attracted the attention of the scientific community for a while. Since autonomous vehicles will directly impact pedestrian behavior at crosswalks, it is necessary to create communication protocols between the two that help build trust in the upcoming technology. We studied the effects of different communication paradigms focusing on the autonomous vehicle's communication interface and developed an algorithm for the analysis of pedestrian behavioral patterns. We describe how we determined the impact of AV on vulnerable road users (VRUs) in shared spaces by analyzing specific parameters, such as the pose and distance of the pedestrian. Results showed that communication with VRUs based on visual cues is not necessarily required for shared spaces, in which informal traffic rules apply.

5.1 Introduction

Thanks to the advances of technology and development teams, both industrial and academic, the field of intelligent transport systems (ITS) has grown exponentially in the last decade. Within the field of intelligent vehicles, (AV) have attracted particular interest within the research community. Further, AV can be defined as robots capable of performing actions without the intervention of humans [1]. In most cases these actions are based on driving from one point A to another point B. In a controlled environment without any other road user, the implementation of driving from A to B would only require creating control algorithms. However, the ultimate goal of AV is to be able to travel in populated environments where there are both non autonomous vehicles and vulnerable road users such as pedestrians or cyclists (or users that are not protected by a vehicle, nor wear any protective artifacts, i.e. helmet and which are exposed to different external and weather conditions [2]). Therefore, the algorithms behind AV need to make sure that the intentions of all road users are understood by all agents in the environment so that the appropriate action is executed in each case.

In conventional situations in which vehicles are operated by humans, pedestrians assume that drivers have no intention of colliding with each other [3] and drivers assume that pedestrians perceive the danger posed by the vehicle and act accordingly. Even so, there are cases where drivers interact with pedestrians transmitting visual or auditory signals to indicate to the pedestrians whether they can cross or not.

In line with this, there are several studies that focus on determining pedestrian behavior patterns at pedestrian crossings with manually operated vehicles. An example is the work developed by the authors of [4], which sought to determine the parameters that affect pedestrians at crosswalks from the perspective of both pedestrian and driver. Results showed that the main factor that affects pedestrian behavior is the distance between them and the vehicle, a finding which was corroborated by the results presented in [5]. In another approach, the authors in [6] studied pedestrian behavior by assessing the location of the vehicle and measuring the frequency at which the pedestrian diverted their gaze from the road to the vehicle. The results showed a higher frequency among the pedestrians at the edges and the middle of the crossing.

When referring to AV, patterns regarding complexity and risk need to be assessed by judging and anticipating the actions of the different actors in the system and determine the rules for coexistence [7]. Similarly, protocols must be created that allow interaction between AV and vulnerable road users (VRUs) to generate a degree of confidence equal to manually driven vehicles [8].

When it comes to the question of trust in AV versus conventional manually operated vehicles, studies such as [9] and [10] provide an insight into the exact protocols that exist that help create the foundation of trust between pedestrians and drivers. In this study, the authors found that pedestrian comfort increased when there was a reaction to the pedestrian observable in a change in the way the vehicle was operated (slight speed change, headlights flashed) or from the driver themselves (hand wave). In an additional study from the American League of Cyclists [11] a perceived increase of risk from AV due to the absence of eye contact was found among pedestrians.

These factors such as perceived response from the vehicle/driver and eye contact between pedestrian and driver provide an insight into building a more trustworthy interaction between AV and pedestrians, particularly at the level of the AV interface. In [12] the behavior of pedestrians exposed to an autonomous vehicle with two different interfaces was studied in a controlled environment. The first interface was based on a strip of light-emitting diode that flashed when the vehicle was going to turn and the second interface consisted of images projected by the AV onto the road surface in front or near the vehicle that indicated if the vehicle was going to turn, accelerate or decelerate. In this work they concluded that for the studied vehicle maneuver (turning), the images helped the pedestrian to decide whether to cross or not, but they did not find any difference in the level of confidence, acceptance and the perceived AV intelligence between the two types of interfaces.

In the same line of research several interfaces that indicated to the pedestrian whether they could cross or not were compared in [13]. The work concluded that behavioral patterns depended mainly on the distance between vehicles and pedestrians. Several other similar experiments have been made to evaluate more advanced communication protocols and interfaces [14], such as artificial eyes that followed pedestrians [15] or implicit forms of communication such as motion patterns of the vehicle [16]. Furthermore, the authors in [17] studied not only different communication interfaces, but also the effects that vehicle size and interface display timing had on the reaction of pedestrians.

Although these studies showed the advantages of using a communication interface between pedestrians and AVs, there are studies that demonstrated that interaction interfaces were not instrumental in a pedestrian's decision to cross. Moreover, the authors of [18] and [19] state that the reaction of pedestrians depend more on the distance between the vehicle and the pedestrian than on the type of interface presented by the AV.

All of the studies mentioned earlier investigated the reaction of pedestrians to different interfaces in different ways. However, most of them were based on qualitative data, simulations, or OZ paradigms in which subjects in the experiment believed that the autonomy was real although the vehicle was operated by a hidden human driver. Although the study of interaction interfaces between pedestrians and AV is ongoing and extensive, the number of studies using actual AV in real situations is quite limited.

The study, classification and, in some cases, prediction of the actions taken by pedestrians is a widely researched subject in ITS. There are studies such as those of [20], [21], and [22] that attempt to classify the actions of pedestrians in video sequences by using neural networks or Markov's Hidden Models. In other works the vehicle crossing situation is modeled as a game where both vehicle and pedestrian are players whose goal is to cross first without colliding with the other player [23].

Although all of these studies present interesting results regarding pedestrian behavior detection in crosswalks, most of them limit their focus to the performance of vehicle control maneuvers in the event of the pedestrian crossing in front of the AV, and/or lack validation in real situations.

To extend the knowledge in the field, in this work we seek to analyze the reaction of pedestrians to an approaching autonomous vehicle without having being informed about an experiment taking place. In doing so, we focus not only on the design and execution of experiments to determine the reaction of pedestrians, but also on the implementation of the most appropriate algorithms to obtain relevant pedestrian behavior patterns in real, novel situations. We seek to develop not only an experimental base with which we can analyze pedestrians, but also to develop an analysis framework that can be used in different experimental situations.

We focus on behavior patterns from the pedestrians that are most exposed to the vehicles (are very close and/or directly in front) and therefore more at risk. To this end, we investigate if communication from the side of the AV affects pedestrian crossing behavior and formulate the following hypothesis:

$H0$: *There is no relationship between the measured pedestrian crossing behavior and driverless vehicle communication signals.*

To this end experiments were conducted in spaces that were shared by VRUs and vehicles without traffic lights, road markings, or signs that indicate the right-of-way such that cooperative action is required. This scenario is applicable for cases such as last mile automatic delivery robots [24].

The following sections extend the publications in [25] and [24] and contribute to the current research with the description of the algorithm for performing the pedestrian analysis in Section 5.3.

5.2 Field Test Description

In our study we seek to obtain pedestrian behavioral patterns with regard to AV that are equipped with communication interfaces to interact with pedestrians. Behavioral patterns are deduced from the pedestrian's body pose at the moment of crossing. In this way, we can classify the intention of pedestrians to cross or not. The pedestrian's body pose is determined based on the approach described in [25], which we describe in detail in this chapter.

To conduct the experiments we used the autonomous vehicle iCab developed by the Intelligent Systems Laboratory (LSI) from the Universidad Carlos III in Madrid, which is equipped with perception and control sensors and operates without the need of a human driver [26]. In addition to the sensors, the platform has an integrated touch screen that acts as a main communication interface for surrounding pedestrians. The interface displays a message to indicate whether a nearby pedestrian has been detected or not. This message consists of an image which was developed as a node in the Robot Operating System (ROS) framework using C++ as its programming language [27].

The vehicle drove autonomously through the central courtyard toward an intersection that is frequently used by students of the university and local residents of the community, since it connects two main streets. The experiments were carried out for two days.

To analyze behavioral patterns we estimated body and head pose of the pedestrians, the relative distance to the vehicle, vehicle speed, and the time to collision (TTC) while the AV was moving. The data required for this analysis was acquired by the sensors of the AV, which detected obstacles that represented a potential future collision by taking into account the rotation angle of iCab. This information was then sent as a Boolean value through ROS to the interface node, which then displayed an image corresponding to the value.

The acceptance and usability of two different visual paradigms (depicted in Figure 5.1) was evaluated with this method. They are described as follows.

- **Baseline:** No image was shown.
- **Red/Green Color:** Based on conventional traffic signs, red indicated stop/do not cross and green walk/cross.
- **Open/Close Eyes:** These images were designed to indicate to pedestrians whether they had been "seen" (detected) or not.

Over the course of two days we recorded interactions between 135 pedestrians and the vehicle. In order to avoid influencing pedestrian behavior and to emulate normal traffic situations, pedestrians were not made aware of the data collection. Although the vehicle was driving autonomously, there was still an external safety control that allowed the vehicle to be stopped in emergency cases where there was a high probability of collision with pedestrians.

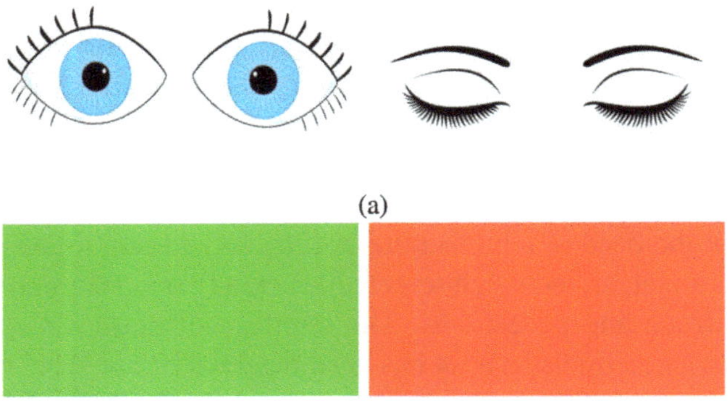

(a)

(b)

Figure 5.1 Images displayed on the communication interface. (a) Open/closed eyes. (b) Green/red color [27].

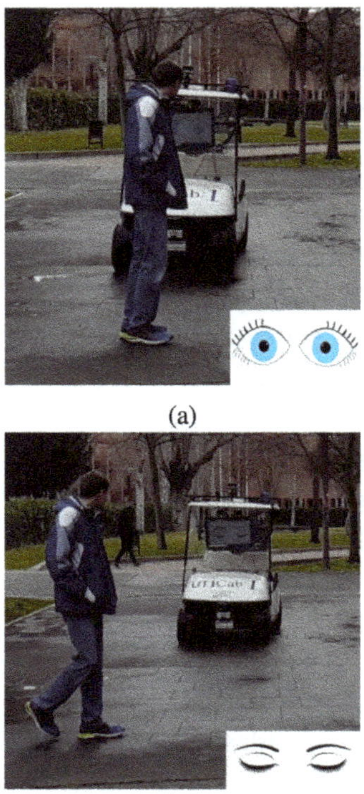

(a)

(b)

Figure 5.2 Example crossing situation where (a) corresponds to the vehicle displaying the open eyes image to pedestrians and (b) the closed eyes image.

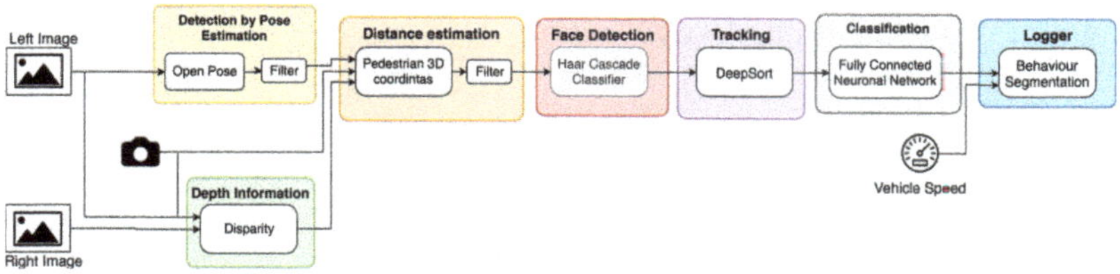

Figure 5.3 Architecture of the analyzing algorithm, inputs being left image, right image, intrinsic information of stereo camera, and vehicle speed. Each module corresponds to a different algorithm used to extract the behavior information of pedestrians.

5.3 Analyzing Algorithm

The study was based on two fundamental parts, the first being the performing of experiments to evaluate the impact of the proposed interfaces on pedestrian behavior. The second part consisted of the design of an algorithm capable of analyzing the interactions so that an automatic analysis of pedestrian reactions could be carried out. To this end we relied on the algorithms previously proposed in [25].

The implemented algorithm used the camera and the speed of the autonomous vehicle to be able to estimate the pose of the pedestrians in the corresponding real coordinates and obtain their position. With this information the behavior was classified into two main categories: crossing or not crossing. The main architecture of this algorithm is depicted in Figure 5.3.

5.3.1 Pedestrian Detection and Pose Estimation

Pedestrian detection and pose estimation was performed using the OpenPose library developed by the Carnegie Mellon University's Perceptual Computing Lab and published in [28] and [29]. This library uses a convolutional, feedback-based neuronal network that creates heatmaps of people's pose keypoints. Moreover, the Part Affinity Fields (PAFs) network obtains a feature representation that preserves the location and orientation of the limbs of the poses. With the PAFs, the library is able to connect the keypoints to obtain an optimal detection of the pose of a person limbs.

With this method it is possible to obtain up to 25 keypoints of the pedestrian pose. These keypoints can be individually extracted according to the pedestrians in a scenario, then used to calculate a Region of Interest (ROI) that is represented by the bounding box of a particular pedestrian in the camera image. Accordingly, we define the upper left corner of this box

(x_0,y_0), width (w) and height (h) as

$$x_0 = min(x_{k_i}) \tag{5.1}$$

$$y_0 = min(y_{k_i}) \tag{5.2}$$

$$w = max(x_{k_i}) - min(x_{k_i}) \tag{5.3}$$

$$h = max(y_{k_i}) - min(y_{k_i}) \tag{5.4}$$

where x_{k_i} is the x coordinate of the keypoint k_i, y_{k_i} the y coordinate in the image and k_i, the i the keypoint of the person's pose where $i \in [0, 24]$.

Although OpenPose allows for a largely accurate estimation of pose keypoints in an image, there are cases in which the network identifies the incorrect pose in areas where there are no pedestrians, or where a pose has too few keypoints or is proportionally not likely to be a walking human. We filtered the results for these errors, discarding poses with less than 20 keypoints or those detections whose bounding boxes width was greater than the height (not likely to be a human pose). This process enabled us to obtain more precise pose readings.

5.3.2 Distance Estimation via Stereo Cameras

To obtain the distance between pedestrians and the autonomous vehicle, the stereo camera of the vehicle was used to calculate the depth information of the image. This type of camera extracts two images of the environment, similar to human eyes, which are separated by a given distance. With these two images it is possible to obtain the disparity image of the environment, which is computed internally by the vehicle using the algorithms of [30].

In order to effectively calculate the depth information of the environment, it is necessary to calibrate the camera so that the images can be rectified. This calibration allows the computation of the camera's intrinsic parameters such as the focal length (f), the base pixels (c'_x,c'_y), and the distance between the cameras (B). Having a rectified image, the disparity and the intrinsic parameters of the camera, the real coordinates of the pedestrian are

$$X_{3D} = \frac{(x - c'_x)B}{d(x, y)} \tag{5.5}$$

$$Y_{3D} = \frac{(y - c'_y)B}{d(x, y)} \tag{5.6}$$

$$Z_{3D} = \frac{fB}{d(x, y)} \tag{5.7}$$

Figure 5.4 Orientation of the (x,y,z) relative distance information from AV shown by the axis in the image.

where (X_{3D}, Y_{3D}, Z_{3D}) are a pedestrian's relative 3D coordinates to the vehicle, following the convention that is shown in Figure 5.4. The representation $d(x, y)$ is the disparity value on pixel (x, y). For simplification of the algorithm, the system was already calibrated before the experiments were performed. Finally, to obtain the position of a pedestrian relative to a vehicle, the 3D coordinates of the pose keypoints are averaged.

It should be noted that at this point, we also filter out and remove detections whose depth Z_{3D} is greater than 15 m. At such a distance, the pose keypoints often present errors, in addition to the fact that for the use case, pedestrians located outside of this range are of no interest, as the vehicle is not an impediment for crossing.

5.3.3 Pedestrian tracking with DeepSort

Unlike in the previous study [25], the tracking algorithm used in this work relied on [31], which is an extension of the Simple Online and Realtime Tracking (SORT) algorithm [32]. In this algorithm a Kalman filter with a state of eight variables is used to predict the location of the bounding box in the next frame from the motion patterns of the detections.

Having a prediction of a pedestrian's position in the following frame and the detections in that frame, the problem arises of matching the predictions to detections, where it is necessary to estimate which tracking belongs to a detection. For this purpose, the authors of DeepSort [33] used the square distance of Mahalanobis, which takes into account the covariance of the distributions, thus allowing a measurement of the distance between them.

At the same time, DeepSort proposes another distance that, together with Mahalanobis, allows the calculation of the relationship between tracks and detections. This second metric is obtained by computing appearance features for the track and the detection and calculating the cosine of the angle between both vectors [34]. The final metric is calculated as follows

$$c_{i,j} = \lambda d^{(1)}(i,j) + (1-\lambda) d^{(2)}(i,j) \tag{5.8}$$

where $c_{i,j}$ is the metric proposed in DeepSort, λ a hyperparameter to associate the cost of the appearance distance $d^{(2)}(i,j)$, and the distance between distributions $d^{(1)}(i,j)$. The appearance features, on the other hand, are obtained using a Convolutional Neuronal Network (CNN) and were trained on a re-identification dataset.

Having the metric between the track and the detection, the Hungarian algorithm is used to make the final match between them. In our case, the detections are captured using OpenPose, unlike DeepSort developers who used a custom neuronal network to detect pedestrians. Using OpenPose detections as an input for DeepSort allow us to track pedestrian with also their pose, maintaining an identification number for each pedestrian through the frames of videos.

5.3.4 Face Detection

To determine whether or not the pedestrian has noticed the autonomous vehicle, we implemented a face detection module by relying on the Haar-Cascade classifier developed by [35], which is based on the determination of the contrast between adjacent features using defined kernels. These features take into account attributes of the face where a defined contrast exists, such as the mouth area and eye area. In this way, pedestrian face representation is the result of using different kernels from different areas of the image.

Haar-Cascade classifies the image by implementing a level detection in different regions of the image. In other words, the classifier works in cascade looking for areas that represent the nose, eyes, cheeks, etc. In the event that an initial attempt bears no results, the classifier discards the area and proceeds with another.

We used the classifier provided by the OpenCV library, which is optimally trained to detect frontal faces. In the context of our study, a frontal face corresponds to pedestrians at least having the vehicle in their field of view.

Although the classifier detects the faces of people watching the vehicle, it also obtained false positives, which were filtered using the points of the OpenPose head pose to remove all face detections that did not have the head pose point.

5.3.5 Velocity

With respect to the speed of the vehicle, the iCab AV has an optical encoder that senses the speed of the wheels, which along with the physical dimensions of the vehicle, allows for a calculation of the vehicle's velocity. The speed of the vehicle was transmitted using a ROS package through the topic $/icab1/velocity\_absolute$ to which the algorithm was subscribed.

5.3.6 Classification

As previously mentioned, the pedestrian pose was represented with 25 keypoints, which were classified using a fully connected neuronal network into "crossing" or "not crossing." A prediction of the intention of the pedestrian to cross or not was not part of this work.

The classifier is based on a neural network with eight fully connected layers, and a binary output layer for whether the pedestrian is walking or not walking. Since the 25 keypoints obtained from the OpenPose library consist of the (x, y) coordinates of each keypoint in the image and the probability p of each point, the network input is a vector $(1{,}75)$. However, before we can obtain a classification of the pedestrian pose, we must normalize the pose to eliminate the bias of the pedestrian's position in the image. In this case, we normalize the points in the following way

$$\hat{x}_{k_i} = \frac{x_{k_i} - x_0}{w} \tag{5.9}$$

$$\hat{y}_{k_i} = \frac{y_{k_i} - y_0}{h} \tag{5.10}$$

where $\hat{x}_{k_i}$ and $\hat{y}_{k_i}$ are the normalized values of the position (x_{k_i}, y_{k_i}) of the keypoint k_i.

We trained this classifier with the JAAD dataset, which consists of 326 videos of situations where several pedestrians cross while a vehicle is approaching. The decision to choose this dataset is based on the fact that it has situations in various scenarios with different weather situations.

Before we could train the classifier, we had to perform a preprocessing step of the dataset, since in our approach we used the pose keypoints to determine the action of pedestrians. This preprocessing step consisted on running OpenPose on the dataset videos to extract and record all pedestrians pose keypoints with the corresponding dataset action label.

Initially, OpenPose presented errors in the JAAD dataset due to image quality. In order to obtain precise estimates of pedestrian pose, a filter was implemented in a script, which, depending on the extracted poses and bounding boxes annotated in the dataset, discarded pedestrians in a given frame. The filter discarded those poses that presented more than five keypoints beyond the annotated bounding boxes in each frame. In the case that a pose was valid, its keypoints were annotated in a csv file, together with the pedestrian's action and bounding box. Therefore, each sequence has a csv file associated with pedestrian annotations, which will be combined in a final csv file to obtain the labels to train the classifier.

After getting the poses of the JAAD dataset we obtained 11556 poses from where 5962 are for crossing and 5594 for not. To train the model we splitted this dataset into 20% so we have 9244 (4768 crossing/4475 not crossing) poses as training set, and 2310 (1194 crossing/1119) as test set.

The model was trained with a GTX1070 with 200 epochs with a batch size of 300. The accuracy results from training and testing are illustrated in Figure 5.5, having the obtained model an accuracy of 98.76 % on the training set and a 94.59 % on the test set.

5.3.7 Behavior Segmentation

To facilitate communication between all of these algorithms we used ROS. This allows the interconnection of processes that communicate with each other through messages using channels called topics. In our case, all the information recorded by the sensors was encapsulated under ROS, which in turn synchronizes the messages using headers with time stamps in each message.

A ROS package was programmed to extract the recorded images by the stereo camera, the intrinsic parameters information of the camera and the speed of the vehicle, and communicate the modules explained earlier. The

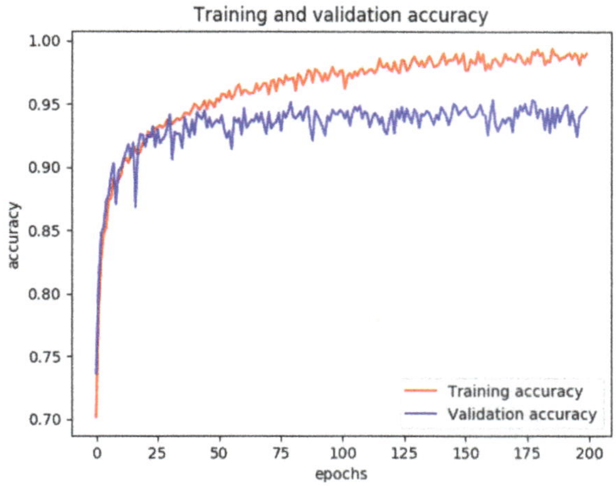

Figure 5.5 Result of the training of the model on JAAD Dataset.

estimation of the disparity occurred in a separate package that send these data to the package.

This all resulted in the estimation of the pedestrian's pose, their face detection, their tracking, and their pose classification by obtaining a state S of each pedestrian, which represented their spacial coordinates, actions, whether or not they had seen the vehicle and the vehicle's speed

$$S = (X_{3D}, Y_{3D}, Z_{3D}, o, a, v) \tag{5.11}$$

where X_{3D}, Y_{3D}, Z_{3D} are the coordinates calculated section 5.3.2, o being the information of whether or not the pedestrian observed the vehicle, a the information of whether or not the pedestrian is walking, and v the speed of the vehicle.

With this state it is possible to determine at any time from the video whether the pedestrian is crossing or not. The action of the pedestrian is initially recorded. Then if the pedestrian is walking, their position with regards to the vehicle determines if a potential collision could occur. In the case the pedestrian is walking, it is resolved that the pedestrian is crossing. If, however, the pedestrian is stopped outside the driving path of the vehicle, it is determined that the pedestrian is not crossing. Pedestrians who are walking but not approaching the path of the vehicle are discarded because they are not interacting with the vehicle closely enough to be exposed to any risk. In this way, at all times we know the distance from the pedestrians to the vehicle, the collision time, whether the pedestrian saw the vehicle, and if the pedestrian crossed or stopped. This information is recorded in a csv file.

5.4 Data Analysis

In order to test the hypothesis H_0, the data acquired by the algorithms explained in section 5.3 were analyzed using SPSS.

As described in section 5.2, the experiments were conducted on three occasions with different methods of interface communication: baseline (no interface), red/green color interface, and open/closed eyes interface. Based on this and the pedestrian information acquired from the implemented algorithms, two main categories were defined:

(1) Pedestrians who visualized the autonomous vehicle and the message and stopped for a moment.
(2) Pedestrians who visualized the autonomous vehicle and the message and continued without apparent change.

In order to test the categorical variables related to the interface shown to the pedestrians, the Pearson χ^2 test was performed. Subsequently, with the data of the distance and collision time between the vehicle and the pedestrians, the unpaired t-test was performed to determine if there was a statistical relationship between these parameters and the interfaces shown to pedestrians.

On the basis of the fact that visual contact plays an important role in unmarked intersections, where the presence of such facilitates cooperative actions while the evasion of visual contact is a way to dominate the other in an interaction [36], [37]; data from the pedestrians who did not visualize the vehicle was also analyzed. Specifically, the average distance and the TTC at which these pedestrians crossed were calculated and compared with the values obtained from the pedestrians who did visualize the autonomous vehicle. The unpaired t-test was also applied to determine if there was a difference in the behavior of these two groups of the pedestrians.

5.5 Results

5.5.1 Algorithm Result

We tested the reliability of the depth information obtained by the stereo camera. For this purpose we compared the pedestrian position estimated by

Table 5.1 Pedestrian behavior depending on the system display condition

Range	Pedestrians	RMSE
0 to 5 m	20	0.17 m
5 a 10 m	20	0.12 m
10 a 15 m	20	0.22 m

the algorithm with the information of the laser that was in the AV. To do this we compute the root mean square error (RMSE) between the information of both sensors by extracting the distance of 60 random pedestrians divided into three groups. The results obtained are depicted in Table 5.1.

With regards to the assessment of the algorithm used to analyze the behavior of the pedestrians, the individual outputs of each module were evaluated frame by frame.

Results related to the estimation of the pose and the detection of the pedestrians using OpenPose, showed a correct detection of 86.58% of the pedestrians that took part in the experiments. From the 13.42% of errors obtained by OpenPose, 10.43% were due to the fact that the pedestrians were in groups far away from the autonomous vehicle. This causes the library for the resolution of the camera to generate erroneous detections of the pedestrians, associating erroneous extremities or generating non-uniform poses.

These results improved when calculating the relative position of the pedestrians with respect to the autonomous vehicle. In this case, the detections whose distance was greater than 15 m were discarded, obtaining a pedestrian detection of 94.76%. This is taking into account that the pedestrians of interest are those relatively close to and in front of the vehicle.

Results regarding the reliability of the depth information obtained by the stereo camera were obtained by comparing the pedestrian position estimated by the algorithm with the information of the laser that was in the AV. To do this we computed the root mean square error (RMSE) between the information of both sensors by extracting the distance of 60 random pedestrians divided into three groups. The results obtained are shown in Table 5.1.

As can be seen, the RMSE for both the Z and X coordinates do not exceed 1 m distance for the three groups, thereby indicating that the use of the stereo camera and the distance calculation algorithm allow for an accurate estimation of the relative position of pedestrians to the autonomous vehicle.

Regarding the implementation of DeepSort, we obtained a tracking of 80.98% of the pedestrians in the different videos. In 19.02% of cases, the

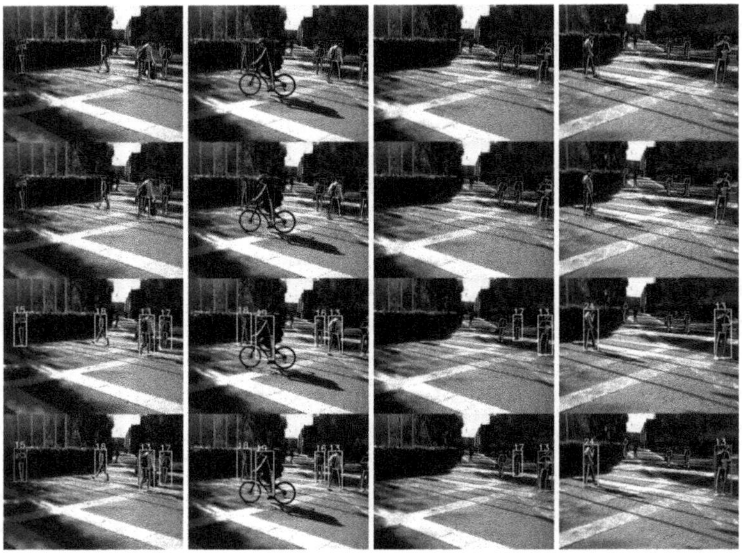

Figure 5.6 Resulting images of each module of the analyzing algorithm. The first row from the top of the image corresponds to the pose and detections obtained with OpenPose. The second row corresponds to the filtered detections taking into account the distance to pedestrians. The third row corresponds to the tracking where the numbers denote the ID of each pedestrian in each frame. The fourth row corresponds to the classification: yellow for pedestrian ID, green for pedestrians that are walking, and red for pedestrians that are stopped.

algorithm lost the pedestrian identification number either because it was identified with another pedestrian or because a new one was assigned since it could not get a match for a pedestrian. Despite the errors obtained by DeepSort, it facilitated the re-identification of the pedestrians along the frames. Error cases were solved by manually noting the cases where the algorithm was wrong and correcting the IDs manually.

Finally, results from the pedestrian pose classification showed an accuracy of 78.67% for all video recordings. These results were due to the fact that the dataset used to generate the neuronal network presented several errors when implementing OpenPose on it. These errors happened because of the quality of the extracted image and the lack of temporal context of the images. In spite of this, the classification allowed us to determine if the pedestrian had crossed in all the cases, since it also takes into account the position of the pedestrians along the entire sequence. The analysis algorithm could therefore classify with an accuracy of 89.34%, whether the pedestrians crossed or not. The individual results of each module are depicted in Figure 5.6, where each row shows images of the individual output of each module.

Table 5.2 Pedestrian behavior depending on the used interface

	Baseline	Green Color	Open Eyes	Red Color	Closed Eyes
Walking	17	9	11	25	21
Stand	3	2	1	1	2

χ^2 test (α =0.05)

		Baseline vs. Green Color		Baseline vs. Open Eyes		Baseline vs. Red Color	
		$(1, N = 31)$	p	$(1, N = 32)$	p	$(1, N = 46)$	p
		1.99	0.158	0.49	0.484	1.77	0.183
		Green Color vs. Open Eyes		Green Color vs. Red Color		Red Color vs. Closed Eyes	
		$(1, N = 23)$	p	$(1, N = 27)$	p	$(1, N = 49)$	p
		0.49	0.484	2.13	0.144	0.50	0.480
Open Eyes vs. Red Color		Open Eyes vs. Closed Eyes		Baseline vs. Closed Eyes		Green Color vs. Closed Eyes	
$(1, N = 38)$	p	$(1, N = 35)$	p	$(1, N = 43)$	p	$(1, N = 34)$	p
0.33	0.57	0.01	0.974	0.41	0.522	0.65	0.420

Table 5.3 Pedestrian distance to the vehicle as well as the TTC at the moment in which they were crossing depending on the type of display showed

Metric	Baseline		Red/Green Color		Opened/Closed Eyes	
	Mean	SD	Mean	SD	Mean	SD
Distance(m)	6.14	3.56	7.38	3.48	6.88	2.76
TTC (s)	7.31	4.64	4.9	6.23	5.10	7.91

t-Test (α =0.05)

Metric	Baseline vs. Red/Green Color		Baseline vs. Opened/Closed Eyes		Red/Green Color vs. Opened/Closed Eyes	
	$t(92)$	p	$t(92)$	p	$t(92)$	p
Distance(m)	1.27	0.20	0.85	0.39	0.67	0.55
TTC (s)	1.51	0.13	1.07	0.28	0.90	0.12

5.5.2 Field Tests Results

Regarding the results from the field test experiments, of the 135 pedestrians 92 (68.17%) looked at the vehicle directly, noticing the message shown on the screen. The rest of the pedestrians (31.86%) walked along the road without looking at the vehicle.

Table 5.2 shows the crossing behavior depending on the interface shown. It presents the results of the Pearson χ^2 test for independent samples for each pair of conditions. The analysis shows that there is no statistically significant relationship between the people who cross or not and the interface shown.

Results related to the distance to the vehicle/time to collision with respect to the image displayed by the vehicle are depicted in Table 5.3.

Table 5.4 Effect of eye contact on interaction with the AV

t-test (α=0.05)

Metric	Without Eye Contact		Eye Contact		t-Test($\alpha = 0.05$)	
	Mean	SD	Mean	SD	$t(133)$	p
Distance (m)	6.93	3.28	7.81	3.56	1.41	0.1599
TTC (s)	5.87	6.71	8.93	12.22	1.37	0.1726

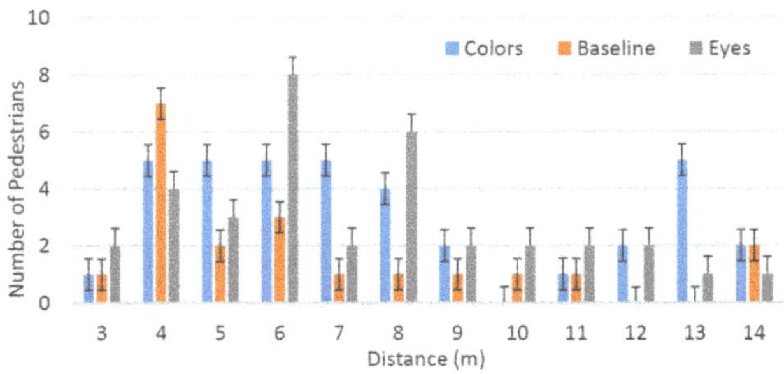

(a)

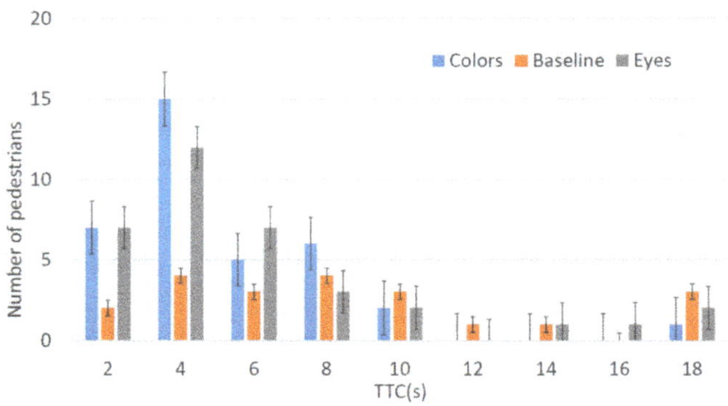

(b)

Figure 5.7 Pedestrian distance to the vehicle (a) and TTC (b) depending on the used interface at the moment of crossing.

The relationship between the image displayed and the distance at which the pedestrians crossed and time to collision was not statistically significant.

Considering these results and the lack of statistical significance in the relationship between the variables, we compared the number of the pedestrians who looked directly at the autonomous vehicle with the number of those who did not and the TTC and distance to the vehicle. Results from this analysis are depicted in Table 5.4.

As it can be observed, the differences were not statistically significant. For both types of the pedestrians the average distance to cross was similar, with a higher TTC for the pedestrians who had visual contact with the autonomous vehicle.

Finally, in Figure 5.7 a bar chart illustrates the distance at which the pedestrians saw the vehicle and crossed, as well as the TTC in relation to the interface shown. From these graphs it can be seen that most of the pedestrians crossed at a distance between 5 and 9 m, in a TTC range of 2 to 8 s.

5.6 Conclusion, Discussion, and Future Work

To begin, it is worth noting that the pedestrians became quickly aware of the AV and recognized the interface that displayed the messages. This was confirmed by comments recorded from persons near the vehicle such as "the car is looking at you" or "wow, this car can see me."

Considering the particular results of the algorithms used, it can be seen that the implementation of OpenPose for the detection has a high fidelity; however, it is necessary to perform processes for filtering the poses due to errors of estimation by the library. This filtering can be initially related to the poses, taking into account the number of keypoints that are not null and the size of the ROI that is calculated with them.

The use of a stereo camera allows to effectively extract the relative position of the pedestrian with respect to the AV for VRUs less than 15 m away. This provides reliable pedestrian detection by using the position as another filtering parameter.

The DeepSort algorithm has a high accuracy in processing the recorded videos of the experiment. However, it is important to note that the errors obtained by the algorithm are due to a Kalmann filter that depends on the movement of the pedestrian, which presents a conflict in images where the camera moves. So it is necessary to calculate the absolute position of the pedestrians with regard to the speed of the autonomous vehicle.

Regarding the classification, although the results obtained allow the classification of the pedestrian poses in most cases, these results can be improved by using neuronal networks that take into account the temporary information of the poses, where not only the current pose is considered, but also the poses of the previous frames. In this way it is not only possible to obtain a more accurate classification, but also to estimate future actions of the pedestrians through prediction algorithms.

The implementation of the modules presented provides a tool to semi-automatically analyze experimental situations where the pedestrians seek to cross while a vehicle approaches. This facilitates these kinds of experiments, since a manual data extraction and processing makes the analysis difficult, inexact, and tedious. With the approach presented here it is only necessary to get the images from the stereo camera of the vehicles and their speed to obtain in a csv file with the actions of the pedestrians.

The results reported regarding the field tests show that the differences among the proportion of the pedestrians who crossed in front of the AV and those who stopped did not depend on the displayed information, as they were not statistically significant. Therefore, we could not reject the null hypothesis.

Furthermore, it was observed that a large number of the pedestrians crossed when the closed-eye or red-eye message was displayed. This confirms that that the type of display did not affect the distance at which the pedestrians crossed or the TTC. This might be due to the settings of the experiment, in which the vehicle could not exceed a velocity of 5 m/s due to safety reasons.

Since we did not obtain a statistically significant relationship between the presence or absence of eye contact and the distance and the TTC, we could not confirm whether eye contact facilitated cooperative actions between the pedestrian and the vehicle in shared spaces,

During the experiments a large number of distracted pedestrians were observed, either using the telephone or conversing. For safety reasons, in many cases where there were distracted pedestrians we had to stop the vehicle, which caused their curiosity. For this reason there is a large number of the pedestrians in the data set who noticed the vehicle only when it stopped suddenly near them.

Finally, from the results presented in Section 5.5, we can conclude that it is not necessary to implement visual communication signals to interact with VRUs in shared spaces where conventional traffic rules do not apply. This confirms the results of [38], [18], and [39] who concluded that information from communication interfaces are not determinant in defining pedestrian behavior, the distance and speed of the vehicle being more decisive. Therefore, future work will explore other types of signals, such as auditory. We will also use sensors that allow the extraction of the 3D information of a wider panorama of the environment, to obtain information from not only in front of the vehicle, but also beside and behind of the AV. With this, we can obtain the amount of the pedestrians who crossed behind the vehicle, for example.

Acknowledgment

This work was supported by the Austrian Ministry for Climate Action, Environment, Energy, Mobility, Innovation and Technology (BMK) Endowed Professorship for Sustainable Transport Logistics 4.0.

References

[1] A. Hussein, F. Garcia, and C. Olaverri-Monreal, "ROS and Unity Based Framework for Intelligent Vehicles Control and Simulation," in *2018 IEEE International Conference on Vehicular Electronics and Safety (ICVES)*. IEEE, 2018, pp. 1–6.

[2] C. Olaverri-Monreal, M. Pichler, G. Krizek, and S. Naumann, "Shadow as route quality parameter in a pedestrian-tailored mobile application," *IEEE Intelligent Transportation Systems Magazine*, vol. 8, no. 4, pp. 15–27, 2016.

[3] A. Millard-Ball, "Pedestrians, Autonomous Vehicles, and Cities," *Journal of Planning Education and Research*, vol. 38, no. 1, pp. 6–12, mar 2018. [Online]. Available: http://journals.sagepub.com/doi/10.1177/0739456X16675674

[4] S. Schmidt and B. Färber, "Pedestrians at the kerb – Recognising the action intentions of humans," *Transportation Research Part F: Traffic Psychology and Behaviour*, vol. 12, no. 4, pp. 300–310, jul 2009. [Online]. Available: https://www.sciencedirect.com/science/article/pii/S1369847809000102

[5] J. A. Oxley, E. Ihsen, B. N. Fildes, J. L. Charlton, and R. H. Day, "Crossing roads safely: An experimental study of age differences in gap selection by pedestrians," *Accident Analysis & Prevention*, vol. 37, no. 5, pp. 962–971, sep 2005. [Online]. Available: https://www.sciencedirect.com/science/article/pii/S0001457505000795

[6] H. Hamaoka, T. Hagiwara, M. Tada, and K. Munehiro, "A Study on the behavior of pedestrians when confirming approach of right/left-turning vehicle while crossing a crosswalk," *IEEE Intelligent Vehicles Symposium, Proceedings*, vol. 10, no. 2011, pp. 99–103, 2013.

[7] A. Allamehzadeh and C. Olaverri-Monreal, "Automatic and manual driving paradigms: Cost-efficient mobile application for the assessment of driver inattentiveness and detection of road conditions," in *2016 IEEE Intelligent Vehicles Symposium (IV)*. IEEE, 2016, pp. 26–31.

[8] A. Hussein, F. Garcia, J. M. Armingol, and C. Olaverri-Monreal, "P2V and V2P communication for Pedestrian warning on the basis of Autonomous Vehicles," in *IEEE International Conference on Intelligent Transportation Systems (ITSC2016)*. IEEE, 2016, pp. 2034–2039.

[9] T. Lagström and V. M. Lundgren, " "AVIP-Autonomous vehiclesínteraction with pedestrians, Chalmers University of Technology.

[10] S. Yang, "Driver behavior impact on pedestrians' crossing experience in the conditionally autonomous driving context," 2017. [Online]. Available: http://kth.diva-portal.org/smash/record.jsf?pid=diva2{%}3A1169360{&}dswid=6775

[11] League of American Byciclist, "Autonomous and Connected Vehicles: Implications for Bicyclists and Pedestrians."

[12] C. G. Burns, L. Oliveira, P. Thomas, S. Iyer, and S. Birrell, "Pedestrian decision-making responses to external human-machine interface designs for autonomous vehicles," in *IEEE Intelligent Vehicles Symposium, Proceedings*, vol. 2019-June. Institute of Electrical and Electronics Engineers Inc., jun 2019, pp. 70–75.

[13] M. Matthews, G. V. Chowdhary, and E. Kieson, "Intent Communication between Autonomous Vehicles and Pedestrians," Tech. Rep. [Online]. Available: https://arxiv.org/pdf/1708.07123.pdf

[14] K. Mahadevan, S. Somanath, and E. Sharlin, "Communicating Awareness and Intent in Autonomous Vehicle-Pedestrian Interaction," in *Proceedings of the 2018 CHI Conference on Human Factors in Computing Systems - CHI '18*. New York, New York, USA: ACM Press, 2018, pp. 1–12. [Online]. Available: http://dl.acm.org/citation.cfm?doid=3173574.3174003

[15] C.-M. Chang, K. Toda, D. Sakamoto, and T. Igarashi, "Eyes on a Car: an Interface Design for Communication between an Autonomous Car and a Pedestrian," 2017. [Online]. Available: https://doi.org/10.1145/3122986.3122989

[16] M. Beggiato, C. Witzlack, S. Springer, and J. Krems, "The Right Moment for Braking as Informal Communication Signal Between Automated Vehicles and Pedestrians in Crossing Situations," 2018, pp. 1072–1081. [Online]. Available: http://link.springer.com/10.1007/978-3-319-60441-1{_}101

[17] K. de Clercq, A. Dietrich, J. P. Núñez Velasco, J. de Winter, and R. Happee, "External Human-Machine Interfaces on Automated Vehicles: Effects on Pedestrian Crossing Decisions," *Human Factors*, vol. 61, no. 8, pp. 1353–1370, dec 2019.

[18] M. Clamann, M. Aubert, and M. L. Cummings, "Evaluation of Vehicle-to-Pedestrian Communication Displays for Autonomous Vehicles," 2017. [Online]. Available: https://trid.trb.org/view.aspx?id=1437891

[19] A. Rasouli and J. K. Tsotsos, "Autonomous Vehicles That Interact With Pedestrians: A Survey of Theory and Practice," *IEEE Transactions on Intelligent Transportation Systems*, pp. 1–19, 2019. [Online]. Available: https://ieeexplore.ieee.org/document/8667866/

[20] H. Zhan, Y. Liu, Z. Cui, and H. Cheng, "Pedestrian Detection and Behavior Recognition Based on Vision," in *2019 IEEE Intelligent Transportation Systems Conference (ITSC)*. IEEE, oct 2019, pp. 771–776. [Online]. Available: https://ieeexplore.ieee.org/document/8917264/

[21] A. Toytziaridis, P. Falcone, and J. Sjoberg, "A Data-driven Markovian Framework for Multi-agent Pedestrian Collision Risk Prediction," in *2019 IEEE Intelligent Transportation Systems Conference (ITSC)*. IEEE, oct 2019, pp. 777–782. [Online]. Available: https://ieeexplore.ieee.org/document/8917142/

[22] D. Ludl, T. Gulde, and C. Curio, "Simple yet efficient real-time pose-based action recognition," apr 2019. [Online]. Available: http://arxiv.org/abs/1904.09140

[23] F. Camara, N. Merat, and C. W. Fox, "A heuristic model for pedestrian intention estimation," in *2019 IEEE Intelligent Transportation Systems Conference (ITSC)*. IEEE, oct 2019, pp. 3708–3713. [Online]. Available: https://ieeexplore.ieee.org/document/8917195/

[24] W. M. Alvarez, M. Angel de Miguel, F. Garcia, and C. Olaverri-Monreal, "Response of Vulnerable Road Users to Visual Information from Autonomous Vehicles in Shared Spaces," in *2019 IEEE Intelligent Transportation Systems Conference (ITSC)*. IEEE, oct 2019, pp. 3714–3719. [Online]. Available: https://ieeexplore.ieee.org/document/8917501/

[25] W. Morales-Álvarez, M. J. Gómez-Silva, G. Fernández-López, F. García-Fernández, and C. Olaverri-Monreal, "Automatic Analysis of Pedestrian's Body Language in the Interaction with Autonomous Vehicles," *IEEE Intelligent Vehicles Symposium, Proceedings*, vol. 2018-June, no. Iv, pp. 1–6, 2018.

[26] D. Gomez, P. Marin-Plaza, A. Hussein, A. Escalera, and J. Armingol, "Ros-based architecture for autonomous intelligent campus automobile (icab)," *UNED Plasencia Revista de Investigacion Universitaria*, vol. 12, pp. 257–272, 2016.

[27] D. M. M.A., D. Fuchshuber, A. Hussein, and C. Olaverri-Monreal, "Perceived Pedestrian Safety: Public Interaction with Driverless Vehicles," in *2019 IEEE Intelligent Vehicles Symposium (IV)*. IEEE, 2019, pp. 1–6.

[28] Z. Cao, T. Simon, S.-E. Wei, and Y. Sheikh, "Realtime Multi-Person 2D Pose Estimation using Part Affinity Fields," in *CVPR*, 2017.

[29] Z. Cao, G. Hidalgo, T. Simon, S.-E. Wei, and Y. Sheikh, "Open Pose: realtime multi-person 2D pose estimation using Part Affinity Fields," in *arXiv preprint arXiv:1812.08008*, 2018.

[30] P. Marin-Plaza, J. Beltran, A. Hussein, B. Musleh, D. Martin, A. de la Escalera, and J. M. Armingol, "Stereo vision-based local occupancy grid map for autonomous navigation in ROS," in *Joint Conference on Computer Vision, Imaging and Computer Graphics Theory and Applications (VISIGRAPP2016)*, vol. 3. SciTePress, 2016, pp. 703–708.

[31] N. Wojke, A. Bewley, and D. Paulus, "Simple online and realtime tracking with a deep association metric," in *Proceedings - International Conference on Image Processing, ICIP*, vol. 2017-Septe. IEEE Computer Society, feb 2018, pp. 3645–3649.

[32] A. Bewley, Z. Ge, L. Ott, F. Ramos, and B. Upcroft, "Simple online and realtime tracking," in *Proceedings - International Conference on Image Processing, ICIP*, vol. 2016-Augus. IEEE Computer Society, aug 2016, pp. 3464–3468.

[33] N. Wojke, A. Bewley, and D. Paulus, "Simple online and realtime tracking with a deep association metric," in *2017 IEEE International Conference on Image Processing (ICIP)*. IEEE, 2017, pp. 3645–3649.

[34] N. Wojke and A. Bewley, "Deep cosine metric learning for person re-identification," in *Proceedings - 2018 IEEE Winter Conference on Applications of Computer Vision, WACV 2018*, vol. 2018-Janua. Institute of Electrical and Electronics Engineers Inc., may 2018, pp. 748–756.

[35] P. Viola and M. Jones, "Rapid object detection using a boosted cascade of simple features," in *Proceedings of the IEEE Computer Society Conference on Computer Vision and Pattern Recognition*, vol. 1, 2001.

[36] T. C. Schelling, *Choice and consequence*. Harvard University Press, 1984.

[37] T. Vanderbilt, *Traffic : why we drive the way we do (and what it says about us)*.

[38] D. Rothenbucher, J. Li, D. Sirkin, B. Mok, and W. Ju, "Ghost driver: A field study investigating the interaction between pedestrians and driverless vehicles," in *2016 25th IEEE International Symposium on Robot and Human Interactive Communication (RO-MAN)*. IEEE, aug 2016,

pp. 795–802. [Online]. Available: http://ieeexplore.ieee.org/document/7745210/

[39] A. Pillai, "School of Science Master's Programme in ICT Innovation Virtual Reality based Study to Analyse Pedestrian Attitude towards Autonomous Vehicles Virtual Reality based Study to Analyse Pedestrian attitude towards Autonomous Vehicles," 2017. [Online]. Available: https://doi.org/10.1145/nnnnnnn.nnnnnnn

6

Intelligent Vehicles and Older Drivers

Joonwoo Son[1,2,*] and Myoungouk Park[1]

[1]HumanLAB, DGIST (Daegu Gyeongbuk Institute of Science and Technology), 42988, Daegu, Technojungang-daero, South Korea
[2]Technical Center, Sonnet.AI, 06764, Seoul, Taebong-ro, South Korea
E-mail: joonwooson@gmail.com

Driving is a complex psychomotor task requiring cognitive, sensory, and physical resources. However, driving-related resources, such as vision, audition, cognition, and physical function, diminish with advancing age. Intelligent vehicles could help older populations by compensating for age-related difficulties. This chapter provides an in-depth review of previous research related to older driver's functional changes and the potential of intelligent vehicles. In this review, we summarize older driver's functional limitations that may increase safety risk, and the possibility of compensating for their limitations through the support of intelligent vehicles. Subsequently, age differences in the acceptance and effectiveness of intelligent warning systems are discussed, based on an on-road experimental study. The results revealed significant age and gender differences, and suggest that it is essential to assess age and gender differences in the effectiveness and acceptance of new in-vehicle technologies designed to help older persons avoid unexpected adverse effects. The final section briefly recommends human–machine interface design considerations for older drivers.

6.1 Introduction

As our global population is aging, maintaining road safety is one of the most serious modern challenges. By 2050, the number of people aged 80 and older is predicted to triple in the Organisation for Economic Co-operation and Development (OECD) countries, and a third of the population will be older than 65 years [1]. Regev et al. [2] indicated that the youngest and oldest drivers have much higher crash risk than drivers of other ages. Among these at-risk age groups, older drivers' increased risk is associated with physical vulnerability and functional limitations in cognition, sensory perception, and physical motor behavior [3, 4]. However, drivers aged over 60 often have high purchasing power; for example, people within this age group purchased 23% of new passenger cars in the United States [5]. As such, automotive designers must understand older drivers' responses to driving demands and age differences in responses to intelligent vehicles. These essential factors could help decide the direction of technology development and policies that may help older drivers compensate for some of their diminished driving capabilities. This chapter reviews older drivers' safety risk and age-related factors that may affect older drivers' capability, how intelligent vehicles can compensate for their deficits, the effects of intelligent warning systems on driver behaviors, and design considerations for older drivers.

6.2 Age-related Limitations in Driving

Age brings with it many capabilities, such as increased wisdom, experience, and knowledge. However, limitations in functions that are related to driving, i.e., vision, audition, cognition, and physical function, also increase in prevalence with advancing age [3]. Despite such age-related limitations, not all older drivers are unsafe because driving judgment improves with experience, which may compensate for diminished capacity [6]. However, judgment may fail with severe consequences in situations with very high momentary mental workload [7, 8]. This section introduces the details of age-related functional limitations from the perspective of driving.

6.2.1 Vision and Audition

Vision is the primary sense utilized in driving; although the percentage of driving-related information that is obtained visually has been subject to debate [10], some researchers have suggested that 90% of such information is obtained through visual input [9]. Adequate visual acuity and field of vision are essential for safe driving, but visual impairment becomes significantly more common with increasing age, as a normal part of the aging process

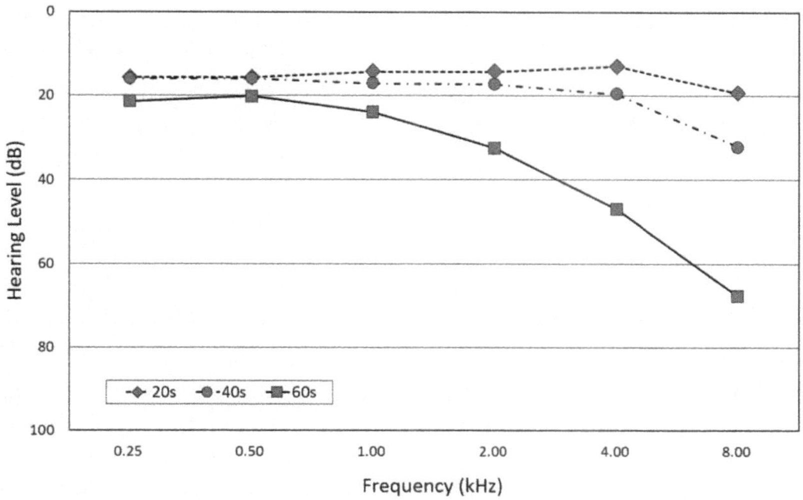

Figure 6.1 Pure tone audiogram results by age groups (from [14]).

[3, 11]. Age-related changes in visual function generally begin from 35 to 45 years, and functional loss in the retina and visual nervous system occurs from 55 to 65 years. These changes lead to issues on peripheral vision, nighttime visual acuity, sensitivity to glare, contrast sensitivity, and color vision [4, 12]. Regarding audition, various estimates have suggested approximately 10% of all middle-aged adults experience significant hearing loss. By age 65, this percentage increases to more than 50% of all men and 30% of all women. Usually, younger adults can detect pure tones with frequencies of up to 15,000 Hz, but older adults have difficulty detecting frequencies above 4,000 Hz [13]. Figure 6.1 shows the decreased ability of the older participants to detect audio frequencies above 2,000 Hz [14].

6.2.2 Cognitive Function

It is common for a driver operating a motor vehicle to engage in many non-driving tasks, such as talking and texting on a cell phone, and operating navigational aids and entertainment systems [15]. Thus, the ability to manage multiple tasks is an essential aspect of safe driving. However, the individual's capacity to manage multiple concurrent tasks generally decreases with age [16, 17]. Older adults generally perform more poorly than younger adults at performing multiple concurrent tasks, such as driving while looking for street signs [13]. Other previous studies have suggested that older drivers are worse than younger drivers at maintaining speed under dual-task conditions [18,19]. The poorer performance of older drivers when engaged in multiple tasks maybe caused by their reductions in divided attention, selective attention, and speed of information processing.

6.2.3 Physical Function

Physical abilities, such as muscle strength, endurance, flexibility, and proprioception, are required to control a vehicle. In general, an older adult may show between 1.5 and 2 times longer respond time than a younger counterpart. The movements of older adults tend to be less precise and more variable than younger adults [13]. Reductions in flexibility, muscle strength, and motor speed as a result of aging or age-related disease are essential factors that may decrease driving ability. Reduced neck rotation may decrease the ability of the driver to turn the head to see relevant stimuli in the periphery, an action that is necessary for safe driving in complex traffic situations [3]. As shown in Figure 6.2, [14] characterized the physical response of drivers by age to different warning sounds. The upper part of the bars represents the accelerator response time, while the lower part indicates the brake response time. The average accelerator response time of older drivers was 170 ms longer than that of drivers in their twenties. In addition to the decreased physical function among older adults, physical fragility may increase the fatality rate for older drivers involved in road traffic accidents [20].

6.3 How Can Intelligent Vehicles Help Older Drivers?

Intelligent Vehicles (IVs) could help alleviate some age-related functional limitations. More specifically, Advanced Driver Assistance Systems (ADAS) can provide useful assistance to older drivers by reducing the difficulties resulting from diminished abilities in motion perception, peripheral vision, and selective attention, and decreased speed of information processing and decision-making [4,21,22]. The weaknesses of older drivers and the potential ADAS features to help address their limitations, as proposed by [4], are summarized in Table 6.1.

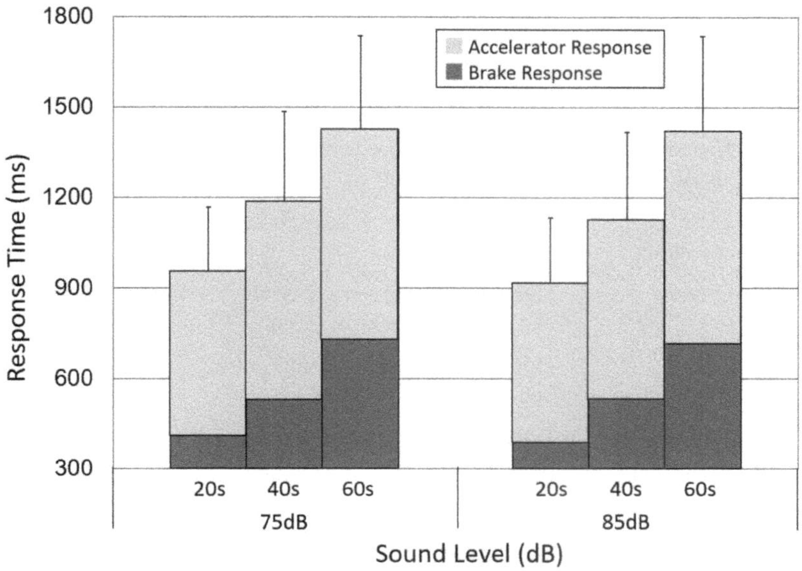

Figure 6.2 Response time on different intensities by age groups (from [14]).

6.4 Intelligent Vehicles and Older Driver

Among the ADAS mentioned earlier, this section introduces the acceptance and effectiveness of forward collision warning (FCW) and lane departure warning (LDW). Many researchers have reported that FCW and LDW systems significantly improve driver safety [23–25]. Concerning older drivers' acceptance of these systems, [26] showed that older adults are willing to pay for new in-vehicle devices and rate the assistance system more highly than do younger drivers. [27] reached the same conclusion based on results of their simulator-based study of the behavioral effects of an in-car tutoring system. The older adults (60–75 years old), as well as the younger drivers (30–45 years old), committed fewer offenses when the system gave feedback messages. Interestingly, while the older drivers were pleased with the warning messages, the younger drivers disliked the system. Son et al. [28] investigated the effects of age on the acceptance and effectiveness of FCW and LDW in an instrumented vehicle. The research methods and findings are summarized in this section. Although this chapter focuses on age difference in the use and utility of IV, it is also important to consider gender differences.

Table 6.1 Weaknesses, difficulties, and ADAS (adopted from [4])

Category	Weakness	Driving Difficulties	ADAS
Sensory Changes	Peripheral vision	Overlooking other road users while merging or changing lanes	BSD[1]
	Nighttime visual acuity	Difficulty seeing pedestrians and other objects at night	NV[2] AFS[3]
	Glare sensitivity	Temporary loss of visual information	HMI[4]
	Contrast sensitivity	Difficulty reading signs & displays, and estimating depth & speed	HMI TSR[5]
	Color vision	Difficulty recognizing similar colors, and reading signs & displays	HMI
	Motion perception	Difficulty judging the movement of road users and their approach speed	FCW[6]
	Hearing	Difficulty recognizing high frequency sounds	HMI
Cognitive Changes	Divided attention	Driving task performance gets worse when performing multiple tasks	LDW[7] HMI
	Selective attention	Overlooking traffic signs and signals	TSR
	Speed of processing	Reaction time increases as the traffic complexity increases	CNS[8] LDW
	Conscious tasks	Difficulty driving in an unfamiliar environment	CNS LDW
Physical Changes	Flexibility head & neck	Overlooking fellow road users when merging or changing lanes	BSD
	dexterity & strength	Difficulty operating on instrument panels	HMI LDW

[1]BSD: Blind Spot Detection
[2]NV: Night Vision
[3]AFS: Adaptive Front-lighting System
[4]HMI: Human–Machine Interface
[5]TSR: Traffic Sign Recognition
[6]FCW: Forward Collision Warning
[7]LDW: Lane Departure Warning
[8]CNS: Car Navigation System

6.4.1 Research Methods

Son et al. [28] conducted a between-subjects single-blind experiment, in which the participants were divided into one group that was supported by the FCW and LDW, and a second group that did not receive this support.

All participants were instructed to drive as similarly as possible to their daily driving style, and no constraints or penalties were used, except that the participants should drive safely. A total of 52 participants were recruited: 26 younger drivers (25–35 years) and 26 young-old drivers (55–65 years). The older group was relatively young, in part to maintain driving safety during the on-road experiments.

6.4.2 Age Differences in the Acceptance of Assistive Technologies

Son et al. [28] reported that the main effect of age on acceptance of the assistive technologies was not significant for both FCW and LDW. However, there was an apparent age-related trend in acceptance of the LDW. The young-old age group reported higher acceptance of the LDW than did younger drivers. This finding is consistent with the results of previous studies, in which older drivers rated the assistance system more highly than did younger drivers [26], and older drivers had a more positive attitude toward the ADAS services than did younger drivers [29]. The acceptance difference between the FCW and LDW may have originated from the difference in effectiveness of the ADAS. The effectiveness differences are discussed in the following section.

6.4.3 Age Differences in Effectiveness of FCW

To analyze the drivers' behavioral changes in response to the FCW system, [28] selected three commonly used measures [23, 25]: the average number of forward collision warnings received (FCWC), the average time headway (TH) when the TH to the closest in-path vehicle was less than 2.5 s, and the percentage of the journey during which the participants were closer than 1.5 s to the closest vehicle (PJ1.5). The effectiveness measures of the FCW system are summarized in Table 6.2. A mixed ANOVA yielded a main effect of age ($p < 0.05$) and a significant interaction between age and gender ($p < 0.05$) on PJ1.5. The effect of age was more pronounced in female participants than in males, as shown in Figure 6.3. The younger participants spent 22.34% of their driving time at headways of less than 1.5 s, as opposed to 17.02% in the young-old group. The interaction between gender and FCW assistance was also significant for average TH. Male drivers maintained higher TH than did female drivers in the FCW-supported condition. Interestingly, female drivers' behavior became much less safe in the FCW-supported condition. As shown in Table 6.2, their

Table 6.2 Results for the effectiveness of the FCW by age and gender (from [28])

	FCW ON			FCW OFF		
	FCWC (n/min)	TH (s)	PJ1.5 (%)	FCWC (n/min)	TH (s)	PJ1.5 (%)
Age						
Younger	0.18 (0.39)	1.36 (0.37)	22.30 (16.46)	0.33 (0.81)	1.35 (0.39)	22.39 (16.05)
Young-old	0.28 (0.76)	1.41 (0.41)	17.70 (15.18)	0.44 (1.33)	1.39 (0.43)	16.34 (15.68)
Gender						
Male	0.23 (0.71)	1.45 (0.40)	18.41 (15.19)	0.33 (1.20)	1.36 (0.40)	19.72 (16.34)
Female	0.23 (0.48)	1.31 (0.37)	21.58 (16.63)	0.44 (1.10)	1.38 (0.43)	19.00 (15.95)

Note: Average with standard deviation in parentheses.

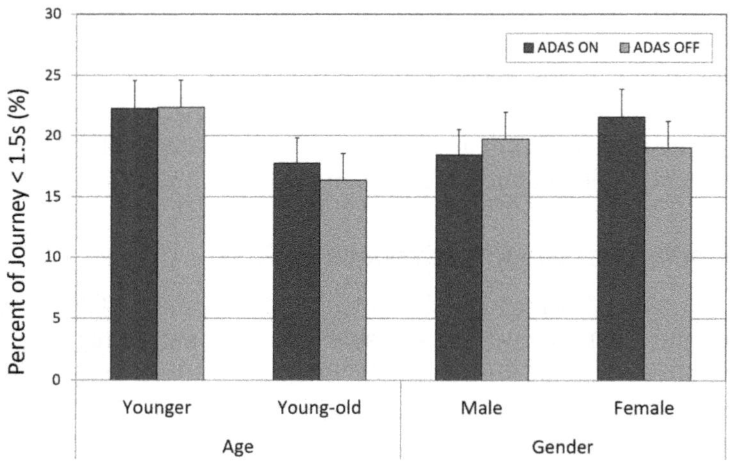

Figure 6.3 Comparison of percent of journey less than 1.5 s by age and gender (from [28]).

Note: Error bars represent the standard error of the mean data.

average time headway decreased by 5.1% to 1.31 s compared with 1.38 s for the non-supported control group of female drivers. This unexpected driving behavior among females could have been due to women's overestimation of speed with increasing speeds [30]. The FCW system used in this study generated a warning sound when the TH was shorter than 1.0 s. The low TH warning criterion may have misled the female drivers to perceive safe TH

Table 6.3 Results for the effectiveness measures of the LDW by age and gender (from [28])

	FCW ON		FCW OFF	
	LDWC (n/min)	SDLP (m)	LDWC (n/min)	SDLP (m)
Age				
Younger	1.38(1.62)	0.30(0.11)	1.79(2.33)	0.28(0.11)
Young-old	1.28(1.56)	0.27(0.10)	1.51(2.05)	0.27(0.09)
Gender				
Male	1.38(1.82)	0.27(0.11)	2.31(2.64)	0.27(0.09)
Female	1.28(1.32)	0.29(0.10)	1.00(1.36)	0.28(0.10)

Note: Average with standard deviation in parentheses.

as shorter, i.e., below 1.5 s, which remained above the warning criterion. As this study found significant age and gender differences in the effectiveness of the FCW, the safety parameters of FCW systems, such as the FCW threshold, should be set according to age and gender characteristics of the driver.

6.4.4 Age Differences in Effectiveness of LDW

The average lane departure warning count (LDWC) and the standard deviation of lane position (SDLP) were used to assess the effectiveness of the LDW [23, 24, 31]. The LDWC represented the number of lane excursions per minute that were executed without activating a turn signal. The SDLP was calculated from the 0.1 Hz high-pass filtered lateral position data, after removing lane changes. A mixed ANOVA indicated that the SDLP was significantly higher for the younger group (M = 0.29 m, SD = 0.11) than for the young-old group (M = 0.27 m, SD=0.09; $p < 0.05$). Although the overall number of LDW decreased when activating LDW (Table 6.3), the main effect of the LDW system on the SDLP was not significant. The results of previous research on the effectiveness of the LDW systems have also been mixed. Blaschke et al. [24] argued that the LDW systems are effective in reducing lane deviations. However, [23] reported no significant decrease in lane deviations, and suggested that the effect of the LDW system on lane deviation was not notable, unless the driver was performing secondary tasks, such as manipulating an infotainment system. The present study is consistent with the results of [23]. In contrast, the current study revealed a significant main effect of age on the SDLP. The younger drivers showed greater lane deviation of 0.29 m compared with 0.27 m in the young-old age group. The age difference in lane deviation could be attributed to the strong correlation between lane deviation and the amount of time while attention is directed

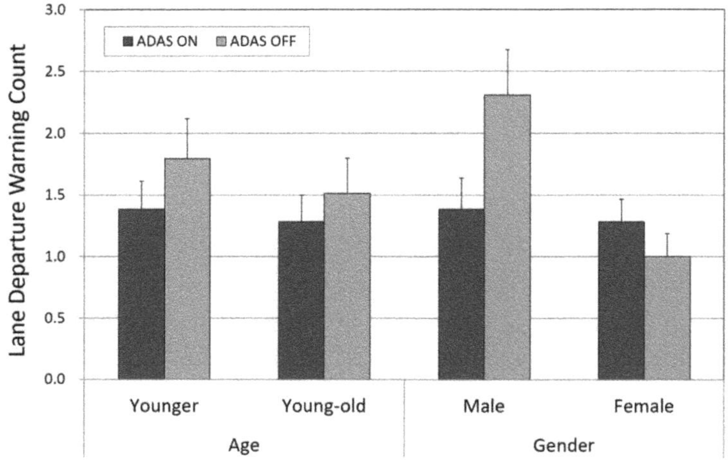

Figure 6.4 Comparison of lane departure warning count by age and gender (from [28]).
Note: Error bars represent the standard error of the mean data.

away from the road [32], and the observation that older drivers tend to check their mirrors less frequently than do younger drivers [33, 34]. The results of the current study also revealed a significant main effect of gender and a trend toward an interaction between gender and the LDW support. The female drivers' number of warnings significantly increased by 28%, to 1.28 times per minute, compared with once per minute in females who did not receive the LDW support (Figure 6.4). This unintended effect could be attributed to gender differences in confidence in their driving skills. Because women have lower such confidence than men [35], they were more easily affected by the LDW warning.

6.5 HMI Design for Older Drivers

6.5.1 Visual HMI Design

As mentioned in Section 1.2.1, older drivers' visual limitations, such as decreased contrast sensitivity and changes in color perception, must be considered in visual HMI design. Regarding lower contrast sensitivity, the contrast of interfaces must be increased in HMI design to maintain visibility of information. When choosing a color, it is important to consider the yellowing of the sclera of the eye, which changes color perception. Since blue light is filtered, blues appear darker. Purple, which is a mixture of red and blue, maybe perceived by older drivers as red [36]. In contrast, older adults are slower to shift attention from one location to another [13]. Thus, older drivers looked at in-vehicle display for a longer duration and less frequently than do younger drivers. In circumstances with relatively high visual demand, older drivers' mean duration of observing information displays often exceeded 1.6 s, while younger drivers' duration remained below 1.6 s [32]. Thus, visual HMI design should present the fewest items possible that must be searched through to perform a task, and remove extraneous information that might capture attention, such as scrolling display elements on a car navigation display.

6.5.2 Audible HMI Design

When designing the audible HMI components for older drivers, it is recommended to use lower frequency sounds, combining with time-varying sounds to convey urgency and criticality. It is also useful to provide a louder option to compensate for older drivers' diminished ability to perceive higher frequency sounds. Specifically, [14] suggested that the frequency of safety warning sounds should be around 3–4 kHz, and the tempo should be around 200 ms. Moreover, the frequency and tempo of other sounds used in cars should be 1 kHz and 500 ms, respectively, to avoid confusing drivers, since many older drivers have difficulties distinguishing safety warning sounds from other general warning sounds.

6.5.3 Multiple-task Design

When younger and older adults are required to perform more than one task at a time, such as driving and watching route guidance, older adults generally performed more poorly than their younger counterparts. The magnitude of the age difference increases with task complexity. However, when tasks are relatively simple, older adults perform as well as young adults. Thus, when designing new HMI operations, it is critical to not require the combined performance of tasks, and to design procedures that are as simple as possible. Regarding interaction types, [37] indicated that older drivers' performance is less affected by cognitive tasks than visual tasks. As a secondary task became more difficult, greater age differences in driving performance were observed. However, older drivers' eye movements and physiological responses were not significantly different from those of younger drivers. The basis of this result could be older drivers' lower awareness of the risks associated with cognitive distraction. This suggests that drivers are not always aware of the detrimental effects on their driving performance of engaging in secondary tasks [38] and often underestimate the risks involved in performing particular tasks [39]. Thus, interactions among age, non-driving task type, and difficulty should be carefully considered when adopting new in-vehicle interfaces and assessing HMI design.

6.6 Conclusions

At the population level, older drivers exhibit a higher fatality rate than younger persons, which is associated with the diminished functional capacity of the former. There is increasing concern regarding older drivers' safety, given the worldwide aging of populations, and IV could be a potential solution to the issue of road safety. The analysis of older drivers' weaknesses and their relevance to road safety suggested that the ADAS systems could compensate for older drivers' limitations. Age differences in the acceptance and effectiveness of the ADAS were reviewed through the results of an on-road experiment. Females and younger drivers showed the lowest acceptance rates, whereas males and young-old drivers were more likely to accept the ADAS systems. From the perspective of effectiveness, the FCW system significantly improved the TH safety margin of the male drivers. However, it also encouraged the TH of the female drivers toward more dangerous driving. The effectiveness of the LDW system was mixed between genders. The male drivers improved their lane departure behavior, while the female drivers, who rated near-lowest acceptance of the system, exhibited the opposite effect. The results suggest that it is essential to consider age and gender differences in the effectiveness and the acceptance of new in-vehicle technologies, to avoid unexpected adverse effects on the driving behaviors of those of particular age and gender. General recommendations for the HMI design for older drivers were briefly summarized in terms of the audible and visual interface, and multiple-task demand.

References

[1] M. Karthaus and M. Falkenstein. Functional changes and driving performance in older drivers: assessment and interventions. *Geriatrics*, 1(2):12, 2016.

[2] S. Regev, J. J. Rolison, and S. Moutari. Crash risk by driver age, gender, and time of day using a new exposure methodology. *Journal of safety research*, 66:131–140, 2018.

[3] K. J. Anstey, J. Wood, S. Lord, and J. G. Walker. Cognitive, sensory and physical factors enabling driving safety in older adults. *Clinical Psychology Review*, 25:45–65, 2005.

[4] R. J. Davidse. Older drivers and ADAS: Which systems improve road safety? *IATSS research*, 30(1):6–20, 2006.

[5] D. Marshall, R. B. Wallace, M. B. Leeds, and J. C. Torner. Enhancing the effectiveness of safety warning systems for older drivers: project report (no. dot hs 811 417). Technical report, United States. Department of Transportation. National Highway Traffic Safety Administration, 2010.

[6] B. Reimer, B. Mehler, J. F. Coughlin, Y. Wang, L. A. D'Ambrosio, N. Roy, J. Long, A. Bell, D. Wood, and J. A. Dusek. A comparison of the effect of a low to moderately demanding cognitive task on simulated driving performance and heart rate in middle aged and young adult drivers. *Paper presented at the annual meeting for the Proc. International Conference on Cyberworlds, IEEE*, pages 493–500, 2008.

[7] L. Hakamies-Blomqvist, S. Mynttinen, M. Backman, and V. Mikkonen. Age-related differences in driving: Are older drivers more serial? *International Journal of Behavioral Development*, 23 (3):575–589, 1999.

[8] L. Harms. Variation in drivers' cognitive load: Effects of driving through village areas and rural junctions. *Ergonomics*, 34(2):151–160, 1991.

[9] G. H. Robinson, D. J. Erickson, G. L. Thurston, and R. L. Clark. Visual search by automobile drivers. *Human Factors*, 14(4):315–323, 1972.

[10] M. Sivak. The information that drivers use: is it indeed 90*Perception*, 25(9):1081–1089, 1996.

[11] R. Klein, Q. Wang, B. E. Klein, S. E. Moss, and S. M. Meuer. The relationship of age-related maculopathy, cataract, and glaucoma to visual acuity. *Investigative Ophthalmology and Visual Science*, 36 (1):182–191, 1995.

[12] G. Haegerstrom-Portnoy, M. E. Schneck, and J. A. Brabyn. Seeing into old age: Vision function beyond acuity. *Optometry and Vision Science*, 76:141–158, 1999.

[13] A. D. Fisk, W. A. Rogers, N. Charness, S. J. Czaja, and J. Sharit. *Design For Older Adults: Principles and Creative Human Factors Approaches*. Human Factors and Aging Series. CRC Press, second edition, 2009.

[14] M. H. Kim, Y. T. Lee, and J. Son. Age-related physical and emotional characteristics to safety warning sounds: design guidelines for intelligent vehicles. *IEEE Transactions on Systems, Man, and Cybernetics, Part C (Applications and Reviews)*, 40(5):592–598, 2010.

[15] A. O. Ferdinand and N. Menachemi. Associations between driving performance and engaging in secondary tasks: A systematic review. *American journal of public health*, 104(3): e39–e48, 2014.
[16] J. McDowd, M. Vercruyssen, and J. E. Birren. *Multiple-Task Performance*, chapter Aging, Divided Attention, and Dual-task Performance, pages 387–414. Taylor and Francis, 1991.
[17] W. A. Rogers and A. D. Fisk. *Handbook of the psychology of aging*, chapter Understanding the role of attention in cognitive aging research, pages 267–287. Elsevier Science, Oxford, 2001.
[18] B. Reimer, B. Mehler, J. F. Coughlin, N. Roy, and J. A. Dusek. The impact of a naturalistic hands-free cellular phone task on heart rate and simulated driving performance in two age groups. *Transportation Research Part F*, 14(1): 13–25, 2011.
[19] J. Son, B. Reimer, B. Mehler, A. E. Pohlmeyer, K. M. Godfrey, J. Orszulak, J. Long, M. H. Kim, Y. T. Lee, and J. F. Coughlin. Age and cross-cultural comparison of drivers' cognitive workload and performance in simulated urban driving. *International Journal of Automotive Technology*, 11 (4):533–539, 2010.
[20] P. Klavora and R. J. Heslegrave. Senior drivers: An overview of problems and intervention strategies. *Journal of Aging and Physical Activity*, 10:322–335, 2002.
[21] C. G. B. Mitchell and S. L. Suen. Its impact on elderly drivers. In *The 13th International Road Federation IRF World Meeting*, Toronto, Canada., 1997.
[22] S. A. Shaheen and D. A. Niemeier. Integrating vehicle design and human factors: minimizing elderly driving constraints. *Transportation Research Part C*, 9(3): 155–174, 2001.
[23] S. A. Birrell, M. Fowkes, and P. A. Jennings. Effect of using an in-vehicle smart driving aid on real-world driver performance. *Intelligent Transportation Systems, IEEE Transactions on*, 15(4):1801–1810, 2014.
[24] C. Blaschke, F. Breyer, B. Färber, J. Freyer, and R. Limbacher. Driver distraction based lane-keeping assistance. *Transportation research part F: traffic psychology and behaviour*, 12(4):288–299, 2009.
[25] A. Ben-Yaakov, M. Maltz, and D. Shinar. Effects of an in-vehicle collision avoidance warning system on short- and long-term driving performance. *H Factors*, 44(2):335–342, 2002.
[26] S. Stevens. The relationship between driver acceptance and system effectiveness in car-based collision warning systems: Evidence of an

overreliance effect in older drivers? *SAE International Journal of Passenger Cars-Electronic and Electrical Systems*, 5(1):114–124, 2012.

[27] D. de Waard, M. van der Hulst, and K. A. Brookhuis. Elderly and young drivers' reaction to an in-car enforcement and tutoring system. *Applied Ergonomics*, 30(2):147–158, 1999.

[28] J. Son, M. Park, and B. B. Park. The effect of age, gender and roadway environment on the acceptance and effectiveness of advanced driver assistance systems. *Transportation research part F: traffic psychology and behaviour*, 31:12–24, 2015.

[29] N. Viborg. Older and younger driver's attitudes toward in-car ITS. *Bulletin*, 181, 1999.

[30] M. Taieb-Maimon and D. Shinar. Minimum and comfortable driving headways: Reality versus perception. *Human Factors: The Journal of the Human Factors and Ergonomics Society*, 43(1):159–172, 2001.

[31] J. Östlund, B. Peters, B. Thorslund, J. Engström, G. Markkula, A. Keinath, D. Horst, S. Juch, S. Mattes, and U. Foehl. Driving performance assessment-methods and metrics. Technical report, EU Deliverable, Adaptive Integrated Driver-Vehicle Interface Project (AIDE), 2005.

[32] J. Son and M. Park. Comparison of younger and older drivers' glance behavior and performance in a driving simulator. In *The 25th Australian Road Research Board Conference*, Perth, Western Australia, Australia, 2012.

[33] C. A. Holland and P. M. A. Rabbitt. The problems of being an older driver: comparing the perceptions of an expert group and older drivers. *Applied ergonomics*, 25(1):17–27, 1994.

[34] H. C. Lee, D. Cameron, and A. H. Lee. Assessing the driving performance of older adult drivers: on-road versus simulated driving. *Accident Analysis and Prevention*, 35(5): 797–803, 2003.

[35] L. A. D'Ambrosio, L. K. Donorfio, J. F. Coughlin, M. Mohyde, and J. Meyer. Gender differences in self-regulation patterns and attitudes toward driving among older adults. *Journal of Women and Aging*, 20(3–4): 265–283, 2008.

[36] M. A. Green. *Roadway human factors: from science to application*. Lawyers and Judges Publishing Company, Inc., Tucson, AZ, 2018.

[37] J. Son and M. Park. Effects of confusion type and difficulty on the behavior and behavior of elderly drivers: visual vs. cognitive. *International Journal of Automotive Technology*, In press.

[38] M. F. Lesch and P. A. Hancock. Driving performance during concurrent cell-phone use: Age drivers aware of their performance decrements? *Accident Analysis and Prevention*, 36(3): 471–480, 2004.

[39] M. P. White, J. R. Eiser, and P. R. Harris. Risk perceptions of mobile phone use while driving. *Risk analysis*, 24(2):323–334, 2004.

7

Integration Model of Multi-Agent Architectures for Data Fusion-Based Active Driving System

Oscar Sipele, Agapito Ledezma and Araceli Sanchis

Departamento de Informática, Universidad Carlos III de Madrid
E-mail: bsipele@inf.uc3m.es; ledezma@inf.uc3m.es; masm@inf.uc3m.es

Most of the time, a pipeline of Extract-Transform-Load (ETL) processes composes a model that constitutes the core of new active safety systems approaches. Nowadays, the inspection of driver behavior during the performance of dynamic driving tasks is included in reasoning models to enhance the ergonomics of active safety systems. Moreover, aspects concerning high automation level such as Take-Over Request (TOR) requires monitoring techniques for driving behavior analysis as well as surrounding prediction to determine the driver's vigilance level. Nonetheless, the integration into Human-In-the-Loop (HITL) driving simulators to assess new models' performance in a controlled environment becomes an arduous task.

This paper presents an architecture model to integrate new active safety systems into the HITL driving simulators, indispensable for the assessment of complex systems whose inputs come from heterogeneous data sources. A mediation engine orchestrates the distributed multiagent architecture, gathering data from information providers and feeding the required inputs for all pipeline components which compose the data fusion based ADAS.

It deployed a support system whose reasoning process merges the driver's face orientation and gaze estimation with driving scene analysis in the STISIM driving simulator. Ten drivers participated in an experimental process based on performing a driving task with sudden and unexpected events to assess the system performance and to measure aspects of human factors.

7.1 Introduction

Data fusion in active driving safety designing is a crucial factor to achieve functional human-centered approach that enhance the driving safety and ergonomics. This imply that the driver's characteristics, needs, capabilities, and limitations command over all aspects of vehicle and environmental design considerations [1]. This is applicable to different improvement and the well-functioning aspects of support system. First, the mitigation of the driver information load for driver support features, especially when the vehicle is equipped with several of them. Second, in vehicles equipped with Automated Driving Systems (ADS), the driver must supervise the performance of driving automation system's engagement to intervene in case of system failure.

Consequently, the monitor must assess several aspects of the drivers' behaviors according to their chosen role during the performance of dynamic driving task (DDT). Stanton et al. stated six relevant factors composing the psychological model which defines the driver's behavior in relation to automated systems operation [2], Thus, the quality of the interaction process initiated through feedback depends on and affects the state of the locus of control, trust, stress, situation awareness, mental model, mental workload, and task demands.

Currently, the driver modeling analyzes physical and physiological aspects to determine signs of deprecated state of the psychological factors cited above. In cases such as models that assess the driver's situation awareness, the complexity of the applied technique is transferred to the deployment in the experimental platform due to the fact that in general, complex models use information from heterogeneous data sources in relation to both the driver and the environment.

Most of the time, the models under development are composed of a mixture of both own- and third-party systems, developed in different platforms and programming languages. Therefore, orchestrating the information flows that feed the subsystems implies an excessive cost. This paper presents the ideas and technical concepts for the deployment of models based on multi-agent architectures that require data fusion techniques. This type of deployment has the main advantage of flexibility because it allows the deployment of a specific test to be completely decoupled from the test platform. This is extremely useful when different experimental processes are being carried out in a driving simulator.

The main idea of this paper is the proposal of a highly scalable integration model whose objective is to achieve an experimental platform that is able to host the functionality of multiple models, running without mutual exclusion and maintaining levels of low dependencies between them.

This paper is an extended version of [3] and, it outlines the technical concepts and insights about the applied designing patterns to deploy a flexible platform for testing data fusion based active driving systems.

The remainder of the paper is organized as follows: Section 2 provides an overview of the background and related work; Section 3 explains the integration model of the proposed architecture; Section 4 describes the experimental setup process for conducting a driving trial with 10 drivers using a driving simulator system; Section 5 exposes obtained results; and Section 6 remarks the conclusions and suggests some future works guidelines.

7.2 Related Work

Data fusion is a crucial aspect of the ergonomics improvement of in-vehicle safety systems. The development of safety systems that involves the process pipelines has been studied in many contexts, including the robustness enhancement of the environment knowledge, the driver's aspects prediction concerning environmental events, and the driver's inspection to assess the taking over request process.

The combination of data sources to enhance the environment knowledge has been studied. Thus, the study conducted in [4] proposed a multiagent architecture to coordinate the information about area profiling with its traffic regulation, obstacles, and vehicle speed for implementing support system features that provide high-level information about specific situations to the driver. The authors in [5] proposed an ontology-based information map constructed though the data gathering about all the existent entities, thus defining a conceptual scene description which allows evaluating the entities' interaction within a context.

Concerning data fusion of the driver's physical aspects and environment information, a study developed in a driving simulation environment proposed a facial analysis approach based on time and frequency features extraction from face landmarks combined with vehicle surrounding information for predicting severe traffic crashes [6]. Other studies build their reasoning model by the correlation of the driver's eye gaze and environment events such as the pedestrian presence to determine the driver's attentiveness [7, 8]. Additionally, prediction of driver's intentions is a relevant research area which involves data fusion. In [9], a cloud-evolving system is applied to determine the driver's actions by the analysis of driving controls and the vehicle dynamics.

The study [10] exposes a probabilistic model which includes the driver's gaze, vehicle dynamics and controls to predict the driver's behavior directly related with the lane changing maneuver. Moreover, several studies combine the driver's physical and physiological aspects to evaluate the driver's wellness. For instance, fuzzy Bayesian networks have been applied to determine the drowsiness by the analysis of variations of the driver's eye state, electrocardiogram (ECG), photoplethysmography (PPG), the skin temperature, and the vehicle speed [11].

Currently, the coming of the fourth automation level arouses different perspectives to assess the driver's state during the take-over-request (TOR). Gold et al. [12] described the driver's take over process as a sequence of several stages which transit from recovering the cognitive information process until acting. In the last studies, the eye gaze movements are used for the driver's attention assessment, sustaining the takeover acceptance procedure after the driver has conducted a non-driving related task [13, 14].

In summary, the work presented in this paper builds on the future perspectives of active driving safety and automation features. The need for solutions that reduce the integration cost of data-fusion based models in driving simulators used as experimental platform is evident. Further, the model validation will able to be improved, for instance, conducting simultaneous performance assessments of different models.

7.3 Deployment Architecture

This section describes the technical details of the proposed integration architecture for the deployment and testing data-fusion models in simulation environments. The proposed architecture encloses the main concepts: firstly, the multiagent paradigm conceptualizes the different models dedicated to analyzing involved aspects in a dynamic driving task. An agent is a distributed system node that executes a machine-learning process to fulfill a stage of a more complex task. Second, a mediation engine (*broker*) holds the interaction between the involved agents and oversees managing and routing the information flows generated and consumed by all system agents.

Figure 7.1 depicts an overview of the involved agents that compose monitors concerning the driver and environment according to information flow directions, whereas the broker establishes the interaction pattern with all the agents involved in the driver-centered safety feature. The agents are classified according to the stages which compose the autonomous driving. The sensoring agents gather the raw data from the driver, the car, and

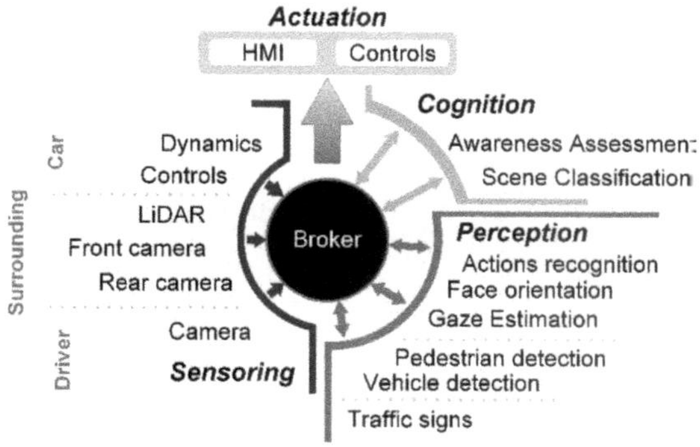

Figure 7.1 Integration model of distributed multi-agent system based on mediation engine.

the environment. In the case of human-in-the-loop driving simulators, the plugged sensors are a physical and emulated devices mixture. The perception agents enclose the machine learning models that use raw data. The cognition agents manage knowledge bases that gather the perception information to assess the DDT performance. Finally, the actuation agents oversee reporting safety feedbacks to the driver or actuate the vehicle control in case of DDT fallback.

As Figure 7.1 illustrated above, the *broker* is a server host that constitutes the central part of the proposed integration model. The *broker* leads a Publish-Subscribe Messaging System as an integration infrastructure, it takes charge of the communication of all agents deployed, whose common goal is to provide a safety feature. Therefore, the *broker* manages the receiving streams from data providers and feeds the subscribers with the information that they require at any moment. Figure 7.2 depicts the publish/subscribe mediation protocol led by the broker. Moreover, it can observe three agent profiles: producers that only write records to a broker's specific topic, consumers that only request records from broker topics, and dual agents that both read and write broker records. Each topic defines the message identifier for the producer's publications, and they are defined for each producer.

The proposed integration model presents several advantages with respect to monolithic integration architectures whereby point-to-point connections establish the communication between different subsystems:

- **Flexibility.** The uncoupled model eliminates the code dependencies, permitting the agent deployment regardless of its implementation.
- **Scalability.** The architecture allows the integration of new agents, incorporating analysis resources that enrich the driving scene definition, and maintaining high throughput.

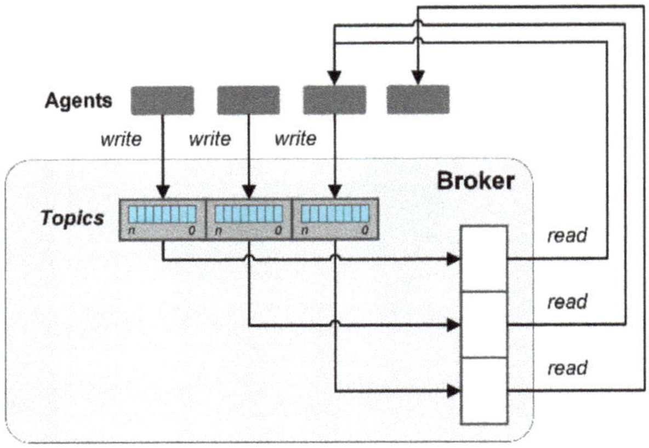

Figure 7.2 Broker architecture integration pattern.

- **Load balancing.** The decentralizing and distribution of processes along different resources stimulate the performance improvement, avoiding the system overload.
- **Support for additional tools.** The architecture can also integrate collecting data tools to consolidate data warehouse for training new models and storing logs for result assessment.

Given the above-mentioned features, the integration architecture allows creating pipelines for composing the safety feature stages, allowing the creation of more complex systems with no dependencies between the different deployed systems.

7.4 Materials and Methods

7.4.1 Materials

Concerning the hardware systems, the *STISIM Drive M300WS* driving simulator system [15] with *VDANL Drive* capability was used. The computing resources were the desktop computers *Intel Core i7-3770* 16 GB Ram with an *NVIDIA GeForce GTX 680* graphics card and, *Intel Core i7-4790K* 16 GB with an *NVIDIA GeForce GTX 960* graphics card, both connected to local area network. As vision device, *Microsoft Kinect v2* sensor was used. Figure 7.3 shows the driving simulation system on execution.

Concerning the software systems, for the implementation and the deployment of the cognitive layer agents, *Java Agent Development Environment (JADE)* was used. Moreover, the cognitive layer was deployed in *Apache*

Figure 7.3 Driving simulator system.

Tomcat 7 Application Server, allowing its connectivity via HTTP protocol. The implementation of agents on driving simulation software was developed using the *STISIM Open Module* capability and .NET Framework. The vision software was implemented using Kinect Development Kit and *EmguCV* computer vision library in C# programming language. The Scenario Definition Language (SDL) provided by the STISIM was used for designing the study cases.

7.4.2 Deployment Details

The architecture deployment on the driving simulation system has been conducted employing different systems implemented in previous works, but this time, these systems were implemented as agents involved on the accomplishment of the functionality requirements of the proposed warning system.

Figure 7.4 shows the decentralized deployment scheme based on the proposed integration model. The support system behind this architecture is integrated by different systems distributed in two computers where each involved system interacts with the broker to publishing or read different topics. The system has as objective to warn the driver in the face of occurrence of five specific cases of use according to their attention over them. In this case, the analysis of driver's visual field determined the driver's attention. The integration of each agent inside the architecture deployment is specified as follows.

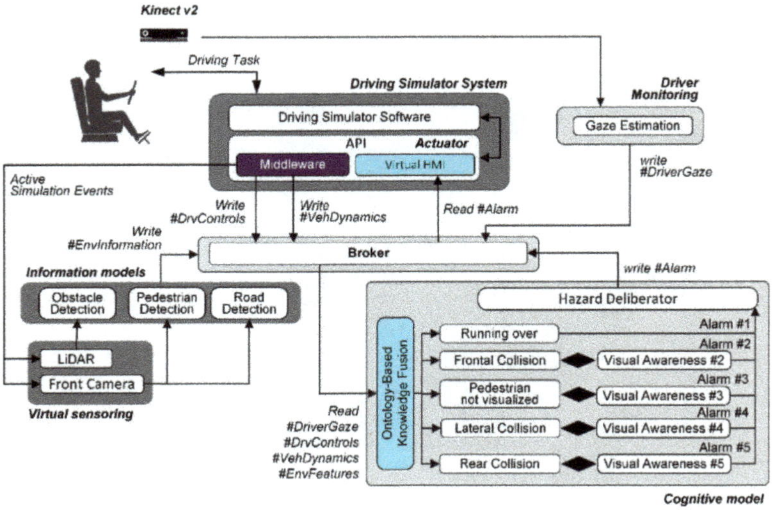

Figure 7.4 Decentralized deployment distributed on simulator system (dark gray) and monitor and reasoning system (light gray).

The agents of the perception and information for the data treatment of the simulated driving environment were implemented. In this work, *LiDAR, Front Camera* were implemented by class model with extraction of the surrounding information accordingly with each technology restrictions. Moreover, the algorithms' behaviors which use those raw data for characterizing the environment (*Obstacle Detection, Pedestrian Detection and Road Detection*) were simulated.

For driver monitoring, the computer vision software was developed using Microsoft Kinect V2 sensor. This system acts as an Information Tier agent (*Gaze Estimation*), and its role is obtaining of the visual area where the driver's attention is focused. Figure 7.5 shows the results of this system, where the driver's pupil position detection (green circle) and the visual area where the driver's attention is focused (red text) can be observed.

The integrated cognitive model was a rule-based alarm system with communication model based on web services [16]. In Figure 7.4, the cognitive model is depicted as a substructure of agents where four kinds of agents can be distinguished:

- The *Ontology-Based Knowledge Fusion* is the process leaded in charge of asking the required information to the broker and structuring of

Figure 7.5 Eye gaze system results: the pupil position (green circle), calibration point (red point), and gaze area estimation (red text).

heterogeneous data. This agent provides an ontological representation of all known aspects involved in the performance of the driving task.

- A set of agents that receive the ontology instance have the aim of identifying the specific cases of use that comprise a driving risk situation (*Running over, Frontal Collision, Pedestrian not Visualized, Lateral Collision, Rear Collision*).
- The identification of each use case activates the Visual Awareness agent actuation. This kind of agents relates the use case progress with the driver's visual perception spread by the Gaze Estimation Agent. Each *Visual Awareness agent* is defined as a parametric model that contains information about where the driver's attention must focus to avoid the incident according to the parameters of the risky driving situation.
- The Hazard Deliberator agent receives the signals from the Visual Awareness agents, in the case of driver's lack of attention has been detected. This agent handles a knowledge base about the priority level of each use case integrated into the system.

This priority level was established arbitrarily according to actors' weakness involved in the driving scene. In the event of driver's inattention on at least two use cases, the Hazard Deliberator agent will take the highest priority hazard, writing its corresponding alarm code in the alarm topic. Finally, the actuator is a *virtual HCI* deployed in the visualization system of the driving simulator [16]. This agent integrates different feedbacks and it reads the topic published by the *Hazard Deliberator* to trigger the appropriate alarm.

Concerning simulation parameters, the simulation frame rate was set to 60 Hz to make easier the synchronization between distributed agents and to avoid the bottlenecks in the slower systems.

7.4.3 Driving Trail Designing

In this subsection, the process conducted for performing a driving trial based on case studies as an evaluation method of the proposed architecture is described. This process is composed of several steps and it has been made following a continuous improvement methodology. As explained later, the cycle of life of this methodology is composed of some activities that can be summarized as PDCA (plan-do-check-act).

The Plan activity consisted on the specification and the designing of case studies. As the starting point, the generic definition of some critical urban traffic situations, such as occluded pedestrians, override risk or frontal collision risk were considered [4]. However, the case studies specification requires the definition of concrete scenes in which these critical driving situations happen and for which the warning ADAS system is necessary. Table 7.1 summarizes the specification of four specific cases as the result of an inductive process with other researchers. Each case study is defined later through its identifier, the location features description, the involved actors, and the actor's action that produces the driving hazard.

In the Do activity were established and tested the parameters of case studies. The primary parameter that should be set was distance with respect to the driver's car at which the case study happens to generate a hazardous, but

Table 7.1 Case STUDIES specification

Case Studies	Description	Scenography
Case 1	Pedestrian occluded by parked cars crosses the roadway.	Residential areas with narrow streets or/and commercial areas.
Case 2	Occluded pedestrian by a car which is waiting for performing a left turning maneuver crosses the roadway.	Residential area with a lot of intersections.
Case 3	Absent-minded pedestrian crosses the roadway from the left or right sidewalk.	At any urban area location.
Case 4	Parked car merges into traffic improperly.	High volume of traffic (i.e., commercial areas and midtown)

surmountable situation. For validating the correct outcome of the designed study cases, each one of them was tested individually on the driving simulation system.

The Check activity consisted on the assessment of the designed scenarios on the driving simulator system. For the evaluation of the driver's behavior on the designed scenarios, some persons had to face the driving challenge. The evaluation with a small group of drivers revealed an aspect of the driver's attention for this experiment. Most of the drivers reflected an excess of concentration when they drove in the simulator system, affecting the experiment progress negatively. As mentioned earlier, the alarm system based on the proposed architecture evaluates the driver's perception toward a specific hazard. Consequently, a drivers' excess of concentration avoids the correct assessment of the system.

At the Act activity were devised a plan for solving the underlying issue of this experiment. As a result of this process, it mandated the need for introducing a mechanism in the designed scenarios for dispersing the driver's concentration enough to test the alarm system in a divided attention way. Hence, for correcting this issue, some divided attention events where included in the driving scenarios. The challenge consists in that, during the driving task, the driver must press the corresponding button from the dashboard when a visual and sound notification appears.

Experimental Setup

Totally, 10 men and women with ages from 21 to 32 (24.5 $\pm$ 3.13), with driving experience of more than 2 years and, between 6000 and 15000 km/year, performed the driving trial. The experiment consists of two sessions performed on separate days.

At the first session, the dated participants give information for the study about their driving skills. Then, a driving task of around 5 minutes is performed by the participant as adaption period to the driving simulator system. After that, each driver performed the driving task in the designed scenario which includes the study cases and the divided attention challenge. Finally, for collecting information about the user experience, the participants answered a brief questionnaire about how they perceived their reaction time.

At the second session, with the implemented alarm system activated, each participant faced with the driving challenge in the other designed scenario. This scenario contains the equal number of situations with respect to the first session scenario. However, driving situations happen in distinct locations and urban scenes. Finally, the participant answered a brief questionnaire about the experience with the alarm system.

7.5 Results

For assessing the proposed architecture, in the experimental process was collected qualitative and quantitative information about its performance.

The qualitative information was obtained from the questionnaires to gather aspects only that are appreciated by the drivers. Table 7.2(a) shows a comparison between information about the driver's reaction time appreciation in the study cases 2 and 3, in absence and presence of the warning system. In this table we can notice that all the drivers could not react with time against these study cases in the absence of the warning system. The study cases 2 and 3 were selected because they are the critical cases in which the driver's inattention can have the most tragic consequences. In the presence of the warning system, the 95% of the answers reflect the impression of reacting either just in time or with time against both study cases.

Table 7.2(b) shows the results of drivers' system evaluation, having as qualitative measures the system utility and the distraction level generated by the system. In this table we can observe that the 80% of the drivers assess the system usability between 4 and 5. About the distraction produced by the system, the 40% of the drivers felt that had been disturbed by the warning system during the driving task performance. The reason is that the tested system does not consider the use case progress when it is identified and whether the driver has overcome it or not.

The quantitative information obtained by objective factors in the face of occurrence of all the study cases is explained later. Figure 7.6 shows the comparative bar-plot with the accident rate happened both in the absence and presence of the warning system, respectively. The subscripts here represent the event occurrences of the same sort of study case. It can be observed the reduction of the number of accidents in the study cases denoted as Case 2, Case 3_1, Case 4_1 and Case 4_2. About the study cases in which the number of accidents increased with alarm system (Case1_1, Case1_2, and Case3_3), the drivers' mental overload could be affected negatively during the driving in the second scenario because they dealt with the driving task, the shared attention, and understanding of the alarm system at the same time. Also, the positioning of these events at the end of the scenario and the previous factor explained earlier suggest the apparition of tiredness signs due to the performance of three simultaneous tasks for 5 minutes.

Table 7.2 Qualitative metrics as result of driver questionnaires (a) Reaction time appreciation for study cases 2 and 3 (b) Overall active safety system evaluation.

(a)

	Without System			With System		
	Reaction Time Appreciation			Reaction Time Appreciation		
Case Study	No Time	Just in Time	With Time	No Time	Just in Time	With Time
Case 2	60%	40%	0%	0%	50%	50%
Case 3	30%	70%	0%	10%	50%	40%

(b)

	Value				
Qualitative Measure	1	2	3	4	5
Utility	0%	0%	20%	50%	30%
Distracting	20%	30%	10%	40%	0%

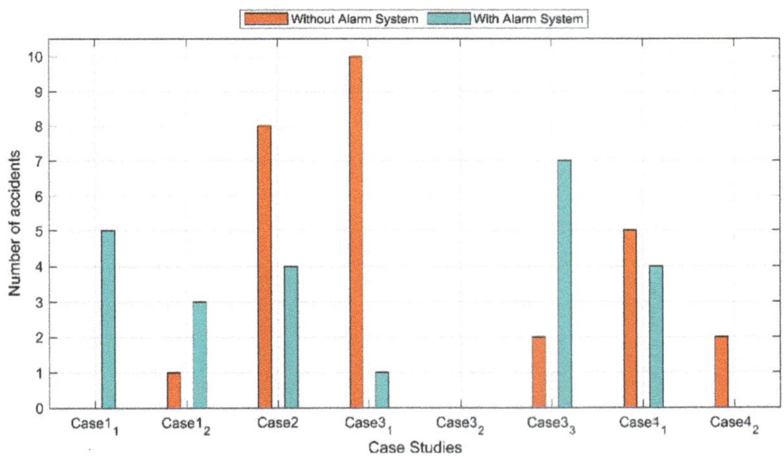

Figure 7.6 Global results in terms of number of accidents.

Figure 7.7 shows a comparative box-plot between the reaction time experimented by the drivers in absence and presence of the warning system based on the proposed architecture. The reaction time was measured from the event that was activated until the driver reacted with a specific action on the driving controls, pressing the brake pedal most of the times. In this plot we can notice the reduction of the reaction time at most times in terms of average and dispersion. For the case study denoted as Case 3_1, we can observe an increase of the measures, although the number of accidents for this case is mainly mitigated. Therefore, this result reflects an early warning report by the system before the events have been triggered.

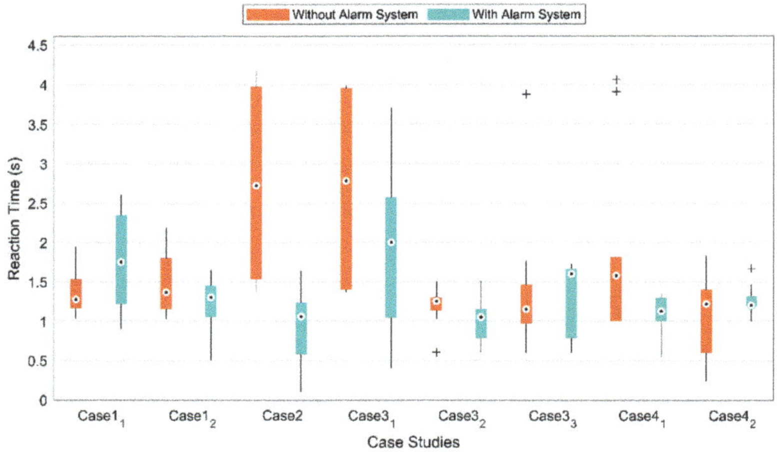

Figure 7.7 Comparative of reaction time measure along the study cases.

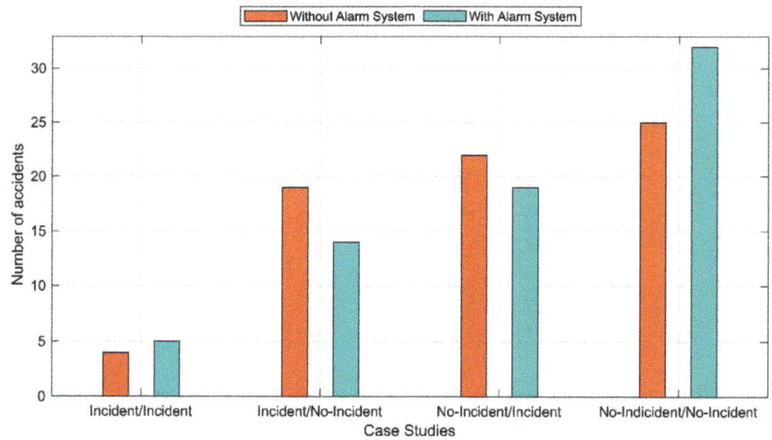

Figure 7.8 Comparative global accidents rate regarding its precedent incident case.

To evaluate the driving trials designing, information has been extracted about the dependency between the consecutive events. Thus, Figure 7.8 shows in general terms how the driver is affected in having to cope up with the event. As it can be observed, there is a slight difference between the different posibilities. Figure 7.9 shows the elapsed time between consecutive events regarding the casuistry about what happened. In general, the driving scenario with the alarm system has a lower elapsed time in most of the cases and, for the No-Incident/No-Incident case, even though it has greater elapsed time, the

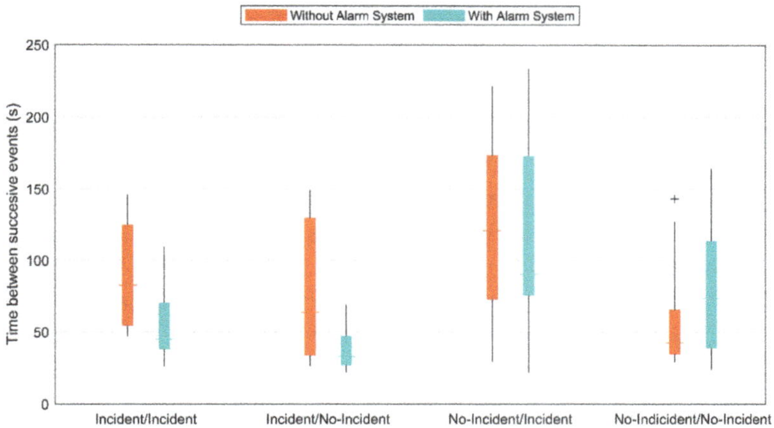

Figure 7.9 Comparative box-plot of obtained time between the consecutive events regarding its precedent event.

global statistics shows a higher numbers of incidents than the driving scenario conducted without the alarm system.

7.6 Discussion

Data fusion is a crucial aspect in the designing of active safety systems, and it joins the present and the future perspectives of human-centered approaches. However, on driving simulator environments, the deployment of model based on fusion data is an arduous task.

In this work has been designed a decoupled architecture oriented to data fusion in which intervene the multiagent systems and Publish-Subscribe Messaging System. As a proof of concept, active safety system whose application gathers information both from the driver's monitor and from environment to determine the driver's attention was deployed on a decentralized basis. A driving trail was conducted for measuring the driver's reaction time and their driving experience, both with and without the deployed support system.

The integration model presents a low latency on the communication across the differently deployed agents. The result of our study evidences the attenuation of the average and standard deviation of the drivers' reaction time in the presented cases of use, justifying both the usefulness of the tested warning system and the proposed architecture in this work. Regarding the results obtained from the number of accidents, in some circumstances, the use cases immediacy makes impossible the driver 's timely reaction; however, the utility of hierarchized ADAS systems leaded by the analysis of the driver's visual perception is also reflected.

As limitations of the presented work, the first approach of the integration architecture was performed by using systems with continuous improvement cycles of life. For this reason, experimental process was bounded to some specific cases of use and promising results were obtained.

In future works, it will include different analysis models such as stress detector and maneuver detection to increase the driver modeling perspective. In addition, it will study other methods to achieve an improved driving simulator experience in terms of visual distraction. Furthermore, the integration model will be tested in another driving simulator system to validate the proposed approach across different platforms.

Acknowledgment

The CAOS Research Group thanks the people who collaborated in the experimental setup. The Spanish Ministry of Economy, Industry and Competitiveness under PID2019-104793RB-C31 and TRA2016-78886-C3-1-R Projects has supported this work.

References

[1] I. Oppenheim and D. Shinar, "Human factors and ergonomics," in *Handbook of traffic psychology*, Elsevier, 2011, pp. 193–211.

[2] N. A. Stanton, M. S. Young, and G. H. Walker, "The psychology of driving automation: a discussion with Professor Don Norman," 2007.

[3] O. Sipele, V. Zamora, A. Ledezma, and A. Sanchis, "Advanced Driver's Alarms System through Multi-agent Paradigm," in *2018 3rd IEEE International Conference on Intelligent Transportation Engineering, ICITE 2018*, 2018, pp. 269–275.

[4] J.-P. Barthès, P. Bonnifait, and P. B. Multi, "Multi-Agent Active Interaction with Driving Assistance Systems," 2010.

[5] A. Armand, D. Filliat, and J. Ibanez-Guzman, "Ontology-based context awareness for driving assistance systems," *IEEE Intell. Veh. Symp. Proc.*, no. Iv, pp. 227–233, 2014.

[6] M. Jabon, J. Bailenson, E. Pontikakis, L. Takayama, and C. Nass, "Facial expression analysis for predicting unsafe driving behavior," *IEEE Pervasive Comput.*, vol. 10, no. 4, pp. 84–95, 2011.

[7] A. Allamehzadeh and C. Olaverri-Monreal, "Automatic and manual driving paradigms: Cost-efficient mobile application for the assessment of driver inattentiveness and detection of road conditions," in *IEEE Intelligent Vehicles Symposium, Proceedings*, 2016, vol. 2016-Augus, pp. 26–31.

[8] L. Fletcher and A. Zelinsky, "Driver inattention detection based on eye gaze–road event correlation," *Int. J. Rob. Res.*, vol. 28, no. 6, pp. 774–801, Jun. 2009.

[9] I. Škrjanc, G. Andonovski, A. Ledezma, O. Sipele, J. A. Iglesias, and A. Sanchis, "Evolving cloud-based system for the recognition of drivers' actions," *Expert Syst. Appl.*, vol. 99, pp. 231–238, Jun. 2018.

[10] V. Leonhardt, T. Pech, and G. Wanielik, "Data fusion and assessment for maneuver prediction including driving situation and driver behavior," *FUSION 2016 – 19th Int. Conf. Inf. Fusion, Proc.*, pp. 1702–1708, 2016.

[11] B. G. Lee and W. Y. Chung, "A smartphone-based driver safety monitoring system using data fusion," *Sensors (Switzerland)*, vol. 12, no. 12, pp. 17536–17552, 2012.

[12] C. Gold, D. Damböck, L. Lorenz, and K. Bengler, "'Take over!' How long does it take to get the driver back into the loop?," *Proc. Hum. Factors Ergon. Soc. Annu. Meet.*, vol. 57, no. 1, pp. 1938–1942, Sep. 2013.

[13] Z. Lu, B. Zhang, A. Feldhütter, R. Happee, M. Martens, and J. C. F. De Winter, "Beyond mere take-over requests: The effects of monitoring requests on driver attention, take-over performance, and acceptance," *Transp. Res. Part F Traffic Psychol. Behav.*, vol. 63, pp. 22–37, May 2019.

[14] X. Li, R. Schroeter, A. Rakotonirainy, J. Kuo, and M. G. Lenné, "Effects of different non-driving-related-task display modes on drivers' eye-movement patterns during take-over in an automated vehicle," *Transp. Res. Part F Traffic Psychol. Behav.*, vol. 70, pp. 135–148, Apr. 2020.

[15] STISIM, "M300WS driving simulation system." [Online]. Available: http://stisimdrive.com/wp-content/uploads/2016/08/new-M300WS-datasheet.pdf.

[16] V. Zamora, O. Sipele, A. Ledezma, and A. Sanchis, "Intelligent Agents for Supporting Driving Tasks: An Ontology-based Alarms System," in *3rd International Conference on Vehicle Technology and Intelligent Transport Systems (VEHITS)*, 2017, pp. 165–172.

Index

About the Editors

Prof. Dr. Cristina Olaverri-Monreal graduated with a Master's degree in Computational Linguistics, Computer Science and Phonetics from the Ludwig-Maximilians University (LMU) in Munich and received her PhD in cooperation with BMW. After working for several years internationally in the industry and academia, she currently holds a position as full professor and BMK-endowed chair of sustainable transport logistics 4.0 at the Johannes Kepler University Linz, in Austria. Her research aims at studying solutions for efficient and effective transportation, focusing on minimizing the barrier between users and road systems. To this end, she relies on the automation, wireless communication and sensing technologies that pertain to the field of Intelligent Transportation Systems (ITS). Dr. Olaverri is Vice-president of Educational Activities in the IEEE ITS Society Executive Committee, chair of the IEEE ITS Austrian Chapter and chair of the Technical Activities Committee (TAC) on Human Factors in ITS. This TAC was recognized with the Award "best TAC of the IEEE ITSS" in 2018 and 2019 respectively. In addition, she serves as an associate editor and editorial board member of several journals in the field, including the "IEEE Transactions on Intelligent Transportation Systems" and "IEEE Intelligent Transportation Systems Magazine (ITSM)". She was recently recognized for her dedicated contribution to continuing education in the field of ITS with the 2017 IEEE Educational Activities Board Meritorious Achievement Award in Continuing Education.

Dr. Fernando García-Fernández is Associate Professor at Universidad Carlos III de Madrid where he focuses his researches in Intelligent Vehicles and Intelligent Transportation Systems involving the use of Computer Vision, Sensor Fusion, and Human Factors.During his professional career, he has pursued research excellence by promoting quality research that has led him to achieve important milestones as a researcher, in addition to several awards. His research activity has resulted in publishing more than 80 papers and 4 industrial patents and the opportunity to work in more than 50 projects with internationally recognized industrial and academic partners.His research activity has allowed him to gain recognition in IEEE Intelligent Transportation Systems Society, where he has been serving in the Board of Governors since 2017 (re-elected in 2020). Besides he has taken a role of the General Chair at some conferences of the mentioned society, as well as the Organizer of the flagship conferences of this society (Intelligent Vehicles Symposium, IV and Intelligent Transportation Systems Conference, ITSC). He is also acting as an editor in the IEEE-Intelligent Transportation Systems Magazine and have served as a guest editor of the reputed journals such as: Journal of Advanced Transportation and Applied Sciences.As lecturer, he has given visiting lectures in different international universities such as University of Buffalo, Università degli Studi di Parma, Universidad Pontificia Bolivariana de Medellin, Universidad de las Fuerzas Armadas de Quito (ESPE), Universidad de La Salle at Bogotá, and Universidad Politécnica de Madrid. Moreover, he gives lectures at his host University in Programming, Computer Vision, Data Fusion, Intelligent Transportation Systems and Control Engineering.

Dr. Rosaldo J. F. Rossetti is a senior research fellow and member of the directive board of the Artificial Intelligence and Computer Science Lab, and a faculty member of the executive committee of the Department of Informatics Engineering, at the University of Porto, Portugal. Dr Rossetti served as an elected member of the Board of Governors of the IEEE ITS Society during term 2011–2013, and was a member of the steering committee of the IEEE Smart Cities Initiative from 2013 to 2017. He is currently chair of the Artificial Transportation Systems and Simulation Technical Activities Committee of the IEEE ITS Society, for which he was the recipient of the Best Technical Activity Sub-committee of the IEEE ITS Society Award, in 2017. He is an associate editor of the IEEE Transactions on Intelligent Transportation Systems, and the ITS Department editor of the IEEE Intelligent System Magazine. Besides being a member of IEEE, he is also a member of ACM and of APPIA (the Portuguese AI Society). Dr Rossetti's main research interests include behavioural modelling, social simulation, spatio-temporal data analytics, and machine learning. He focuses on applications of multiagent systems as a modelling metaphor to address issues in artificial transportation systems and simulation, future mobility paradigms and urban smartification, and explores the potential uses of serious games and gamification in transportation and mobility systems. He holds a PhD in Computer Science from INF-UFRGS, Brazil (2002), having carried out his doctoral research at Leeds University's Institute for Transport Studies, UK.

Lightning Source UK Ltd.
Milton Keynes UK
UKHW020846150121
377097UK00001B/27

9 788770 222044